HUMAN ANATOMY AND PHYSIOLOGY

TRABECULAR AND CORTICAL BONE

MORPHOLOGY, BIOMECHANICS AND CLINICAL IMPLICATIONS

HUMAN ANATOMY AND PHYSIOLOGY

Additional books in this series can be found on Nova's website
under the Series tab.

Additional e-books in this series can be found on Nova's website
under the e-book tab.

HUMAN ANATOMY AND PHYSIOLOGY

TRABECULAR AND CORTICAL BONE

MORPHOLOGY, BIOMECHANICS AND CLINICAL IMPLICATIONS

VARSHA SUBRAMANIUM

AND

MANISH BHATNAGAR

EDITORS

NOVA BIOMEDICAL

New York

For permission to use material from this book please contact us:
Telephone 631-231-7269; Fax 631-231-8175
Web Site: http://www.novapublishers.com

Library of Congress Cataloging-in-Publication Data

Library of Congress Control Number: 2013947022

ISBN: 978-1-62808-766-6

Published by Nova Science Publishers, Inc. † New York

Contents

Preface vii

Chapter I Trabecular Bone: Morphology, Biomechanics and Clinical Implications 1
Ebrahim Bani Hassan and Ali Ghasem-Zadeh

Chapter II Cortical Bone As an Enduring Depot of Bone-Seeking Contaminants – Evaluation of Cortical Bone Resorption Rate on the Basis of a Contaminant (^{90}Sr): Measurement in Humans 37
E. I. Tolstykh, N. B. Shagina, M. O. Degteva, L. R. Anspaugh and B. A. Napier

Chapter III Evaluation of Apparent Stress and Strain at the Fracture of the Human Trabecular Bone under Quasi-Static Compression Load Using Micro-CT Finite Element Simulation 77
Ridha Hambli

Chapter IV Application of Computational Modeling to Predict the Effect of Exercise on Whole Bone Strength 115
C. S. Florio

Chapter V Advancements in Cortical Bone Imaging 137
Alexis Donneys, Noah Nelson, Salman Ahsan, Jonathan Peterson, Michael Morris, Francis Esmonde-White, Steven R. Buchman, Stewart Wang and Benjamin Levi

Chapter VI	Morphology of the Pelvis (Acetabulum) and Femur *Masaaki Maruyama, Judy R. Feinberg, William N. Capello and James A. D'Antonio*	**163**
Chapter VII	The Release Kinetics of Bone Morphogenetic Proteins from Demineralized Cortical Bone Grafts *William S. Pietrzak and Saba N. Ali*	**191**
Chapter VIII	Preservation of Cortical Bone in Oral and Maxillofacial Surgery Procedures *Mohammad Hosein Kalantar Motamedi*	**207**
Index		**217**

Preface

In this book, the authors present topical research in the study of the morphology, biomechanics and clinical implications of trabecular and cortical bone. Topics discussed include trabecular bone biomechanics and morphology; cortical bone as an enduring depot of bone-seeking contaminants and the evaluation of cortical bone resportion rate on the basis of a contaminant (90SR) measurement in humans; evaluation of apparent stress and strain at fracture of human trabecular bone under quasi-static compression load using micro-CT finite elements simulation; application of computational modeling to predict the effect of exercise on whole bone strength; advancements in cortical bone imaging; morphology of the pelvis (acetabulum) and femur; the release kinetics of bone morphogenetic proteins from demineralized cortical bone grafts; and preservation of cortical bone in oral and maxillofacial surgery procedures.

Chapter I – The mammalian skeleton is mainly composed of *a.* cortical (compact) bone (CtB), and *b.* trabecular bone (TbB) also known as cancellous and spongy bone. The distribution of these two has evolved to provide ultimate bone strength, flexibility or stiffness where most needed. TbB is usually found in flat bones, cubic/short bones (e.g. of skull, pelvis, vertebrae …) and the extremities of long bones. There is no definitive border between CtB, subchondral bone (ScB; a variant of CtB) and TbB and their borders usually change - regress and progress - in association with aging, loading, disease, growth or nutritional status. As we will explain, this interface is biologically the most active location in the bone, where most of bone modelling and remodelling processes are going on. Anatomically TbB is in close contact and interaction with bone marrow – which lies in the inter-trabecular spaces. Due to its vast surface area, TbB also provides the most accessible source of minerals mid- to long term. General principles of bone resorption and formation also apply to TbB. However remodelling in TbB is predominantly hemi-osteonal (surface erosions or trenches), and hence bone surface area dependent - unlike CtB which is remodelled in osteonal (cylinder) format. Having less apparent density makes TbB lighter and less stiff, but not essentially less tough. The biomechanical properties of TbB are location-dependent and vary from flexible and tough (e.g. ribs) to fairly stiff (e.g. vertebral bodies). These properties are determined partly by the amount (percentage) of CtB and ScB in a particular bone. Biomechanically TbB is fatigue resistant - i.e. resists crack formation and dissemination due to repetitive loading, a feature that partly differentiates it from CtB and ScB - though some fatigue occurs in TbB. On the other hand, TbB is not fracture resistant. Especially in osteoporotic states (where trabeculae are partially or mostly resorbed and the support of the CtB is minimal due to thinning of cortex) fractures do occur commonly. The most well-known examples of these fractures are femoral neck and vertebral body pathologic fractures in the elderly. This chapter is aimed at providing the basic understanding of physio-anatomy and functional structure of TbB and its interaction with ScB, CtB and soft tissues. Also the mechanisms by which pathology initiates and progresses in TbB, with an emphasis on osteoporosis are further discussed.

Chapter II - Cortical bone is a major part of the human skeleton and amounts to about 60-80% of the skeletal mass. The rate of cortical bone remodeling is low and elements incorporated in the crystal structure of cortical bone remain there for a long period of time. As a result, cortical bone becomes a “long-term storage” for different environmental contaminants. In this study accumulation of bone-seeking contaminants is considered by an example of strontium-90 (^{90}Sr) accumulation investigated in long-term studies at the Techa River (Southern Urals, Russia) and studies of global fallout from nuclear weapon tests. Accidental ingestion of significant levels of ^{90}Sr by residents of the Techa Riverside villages (about 30,000 persons) occurred in the 1950s as a result of the river contamination by liquid radioactive wastes. Strontium-90 contents in the skeleton (mainly in cortical part of the skeleton) have been measured *in vivo* in these residents over the last 60 years. In total, more than 40,000 measurements have been performed. Two critical periods in accumulation of bone-seeking contaminants have been found: (1) period of juvenile growth spurt at the age of 12–15 years, when both incorporation and long-term retention of contaminants is higher by a factor of 4–5 compared to mature adults or younger children; and (2) period of pregnancy and lactation, when contaminant accumulation is higher by a factor of 1.5–2 compared to women of reproductive age. Many residents (about 800 persons) were measured for their ^{90}Sr content for three or more times over a long-term time interval (5–24 years); thus, these extensive data allow evaluation of cortical bone-resorption rate, because resorption by osteoclasts is almost the only way for elimination of the contaminant from the cortical bone. Individual estimates of cortical bone-resorption rate have been obtained for each person in the study that included men and women aged 30–85 years. It was found that rate of resorption is independent of age in adult men aged less than 55 years and in adult women before menopause. In this age period, the rate of cortical bone resorption is independent of gender and close to 3%/year. In women after menopause, the rate of cortical bone resorption depends on time since the onset of menopause. The rate increases by a factor of 2–2.5 compared to pre-menopause and remains at this level for 10–12 years; after that, there is a statistically significant decline in the rate of cortical bone resorption. In men after 55 years of age there is a gradual age-dependent increase. Statistically significant negative relationships were also found between cortical bone-resorption rate and ^{90}Sr-body burden or dose absorbed in skeletal tissues sensitive to radiation exposure [the red bone marrow and the cells lining endosteal surfaces]. However the effect of ^{90}Sr-radiation exposure is insignificant for most of the persons included in the study of cortical bone-resorption rates. As a whole, the effect of ^{90}Sr-radiation exposure on the cortical bone-resorption process was less than the effect of age or menopause in ageing persons.

Chapter III - Micro-finite element (micro-FE) analysis is considered nowadays as a standard tool for the assessment of the mechanical properties of human trabecular bone. Most micro-CT FE analysis of this type of bone has focused on linear and non-linear analysis to evaluate bone failure properties. However, prediction of the failure properties of trabecular bone, including the damage process under monotonic excessive load and taking crack initiation and its progressive propagation until complete bone failure into consideration, is still lacking. In the present work, an isotropic micro-CT finite element model at bone tissue level coupled to a damage law was developed in order to simulate the fracture of human trabecular bone specimens under quasi-static compressive load and predict the apparent stress and strain at fracture and the final fracture pattern of the specimens. The element deletion technique was applied in order to simulate the progressive fracturing process of bone tissue. To prevent mesh-dependence that generally affects the damage propagation rate, regularization technique was applied in the current work. The model was first validated with experimental results performed on seven trabecular specimens. It was then applied to assess the apparent stress and strain at fracture of twenty-three human trabecular bone specimens harvested from different bone sites. The simulation results were compared with reported experimental results to investigate the accuracy of the micro-FE model in predicting the apparent failure properties of trabecular bone. It was found that the predicted failure properties agreed very well with the experimental ones.

Chapter IV - As bone adapts to changes in its environment, the loss of bone strength due to age, long term immobilization, or lengthy stays in reduced gravity conditions can lead to an increased risk of fracture. These strength adaptations are a result of phenomena that modify the mechanical properties of the bone. Changes occur both in the mineral content of the bone tissue and in the overall shape of the bone. Pharmaceutical countermeasures to decreased bone strength systemically alter the mechanisms that control the rates and amounts of mineral release from within bone tissue. This limits the reduction of the density of the bone and, thus, limits the weakening of the bone material itself. Rather than targeting the mineral loss mechanism directly, mechanical countermeasures, such as exercise, work independently of the density changes by amassing bone material on the surfaces of the cortical regions of loaded bones, thus altering the overall bone shape. When sufficient levels are reached, the increased size can improve the overall strength of the bone despite the material weakening that occurs with decreased density. While weight bearing exercises are widely accepted methods to offset the loss of bone strength with few adverse side-effects, quantitative comparisons of the benefits of the many different types of such exercises have not been well established through clinical or experimental means. Computational modeling methods have the ability to isolate the effects of individual system parameters, which is often difficult to achieve experimentally. Therefore, computational models may provide a means to objectively compare the benefits of the various suggested bone strengthening exercises. Developed models simulating cortical bone surface adaptation have shown the significant effect of loading parameters on bone shape, indicating that a better understanding of the relationships between the changes in the mechanical environment and the localized strength of a bone may be achieved through computational techniques. Therefore, the use of such computational models may be advantageous in the development of effective exercises for targeted strengthening of bone, particularly in regions that might be especially susceptible to fracture.

Chapter V - Cortical bone is a unique and dynamic tissue due to its complex micro-architecture, structural role in the human skeleton and ability to continuously remodel. Its dense nature makes it clearly distinguishable from other tissues by means of imaging. Through a portrayal of radiodensities, imaging provides both qualitative clues and quantitative metrics reflecting information about cortical integrity, physiology and function. This chapter reviews the imaging of cortical bone from its rudimentary beginnings (X-ray) to current, modern emerging technologies such as Nano-Computed Tomography and Vascular Microangiography. The clinical importance of cortical bone imaging, as well as its value in research and development is emphasized. Cortical bone morphomics is also highlighted as a potential predictive, rapid clinical screening tool in the modern day trauma bay. The authors further discuss how current innovations in cortical bone imaging and corroborating analytical methodologies such as Raman spectroscopy may lead to useful investigative tools to aid in the early detection of pathologies such as Osteoradionecrosis, Bisphosphonate Related Osteonecrosis of the Jaw (BRONJ) and Heterotopic Ossification.

Chapter VI - Introduction: There are gender differences between male and female in morphology of the pelvis and femur. To confirm this finding, we studied the pelves and femurs in Department of Anthropology, Cleveland Natural History of Museum. Materials and Methods: The bones of 50 males and 50 females (total 200 hip joints) were selected from 3,300 skeletons in the museum at random under the following condition: the same distribution of age between the two groups, such as 5 (under 40), 5 (40-49), 15 (50-59), 15 (60-69), 10 (70 or more); and 20% of black in each age class of the groups. The notch acetabular angle was defined as an angle between a line connected the sciatic notch with the posterior acetabular ridge and a line connected both the posterior and anterior acetabular ridge. Each anteversion angle and inclination of the acetabulum, and antetorsion angle of the femur was also measured and statistically analyzed. The results were analyzed by unpaired t-test and considered p values of less than 0.05 to be significant. Femoral configuration, such as antetorsion of the femoral neck, neck shaft angle, medial offset of the femoral head, anterior bowing and lateral bowing of the femoral shaft was also examined in direct measurement and in radiographs.

Results: The shape of the anterior column of the acetabulum was classified in four types, such as straight, curved, angular, and irregular. Most of the pelves have the curved configuration (60.5%). The notch-acetabular angle was 88.4 ± 3.0° in male and 89.6 ± 3.8° in female ($p=0.013$). All of the angles were ranged within 90 ± 10°. The anteversion angle of the acetabulum was 18.5 ± 5.8° in male and 21.3 ± 7.1° in female ($p=0.002$). The antetorsion angle of the femoral neck was 9.8 ± 9.0° in male and 9.8 ± 8.0° in female ($p=0.954$). There were no correlations between the notch acetabular angle and the other angles (r^2<*0.08*). The neck shaft angle of the femur in neutral rotation was 123.2 ± 6.3° in male and 125.3 ± 5.4° in female (*p=0.014*). The neck shaft angle of the femur in derotation was 121.8 ± 6.1° in male and 123.8 ± 5.2° in female (*p=0.0158*). The medial offset of the femoral head in neutral rotation was 47.2 ± 6.4 mm in male and 41.9 ± 5.9 mm in female (*p<0.001*). The medial offset of the femoral head in derotation was 50.1 ± 5.3 mm in male and 44.3 ± 5.4 mm in female (*p<0.001*). The anterior bowing and lateral bowing of the femoral shaft was 10.3 ± 2.0° in male and 9.0 ± 2.4° in female (*p<0.001*).). The lateral bowing of the femoral shaft was 15.7 ± 8.7° in male and 10.1 ± 9.7° in female (*p<0.001*). There were three distinct femoral shaft configurations, such as lateral, double, and medial curve, of femoral shaft in neutral rotation, in anterior-posterior view.

Discussion and Conclusion: Dimensional differences of the pelvis and femur between females and males were significant with a wide variation within sexes. Most of acetabula had a non-straight acetabular ridge. Configuration of the ridge cannot be detected on standard radiographs nor intraoperatively. Measurement of anteversion angle is dependent upon the point of measurement along the acetabular ridge. Anteversion angle could easily be over-estimated in cases of non-straight anterior acetabular ridge. The notch acetabular angle had small standard deviations, such as 3.0° or 3.8°, and an extremely small range of 90 ± 10° in all of the cases. Although female has slightly greater angle than male, the difference was only 1.2°. For the cases with posterior osteophyte of the acetabulum, this angle was decreased but more than 80°. In addition, this angle was not correlated with the other parameters. As a result, the notch acetabular angle was useful to identify the anteversion of acetabulum intraoperatively. Medial offset measurement assumes straight long axis of femur. Standard radiographs of hip include only proximal femur from which the long axis cannot be accurately determined. Offset measurement was influenced by rotation of femur. The actual amount of offset may be under-estimated on pre-operative radiographic assessment. Effect of lateral bowing on the apparent long axis of femur was a change of +3.5° varus. Similarly, bowing of the femur was found to be more complex than previously thought. In addition to anterior bowing, some femurs were found to have lateral and even medial bowing. Dimensional differences of the pelvis and femur between females and males were significant with a wide variation within sexes.

Chapter VII – Demineralized bone matrix (DBM) grafts, created by removing the mineral phase of cortical bone by treatment with acid, are widely used throughout the skeleton for bone repair applications. Bone morphogenetic proteins (BMPs) are bioactive proteins associated with bone matrix that play an essential role in osteoinduction. Although it is believed that BMP release from DBM, in situ, is a required initial step, little is known about the release kinetics. Such knowledge, and an understanding of the parameters that affect this, can potentially lead to improved DBM graft technologies. This chapter examined the long-term release profile of BMP-7 from bovine particulate DBM (106-710 μm particle size). The DBM (1.5 g) was exposed to 40 ml of Sorensen's buffer in closed vessels that underwent continuous mixing. Periodically, at intervals up to 84 days, the BMP-7 content of guanidine extracts of the DBM was measured with Enzyme Linked Immuno Sorbant Assay (ELISA), with ELISA used to measure the BMP-7 content of the buffer directly. The initial concentration of BMP-7 in DBM was 24.6±1.6 ng/g DBM and declined to 13.8±2.16 ng/g DBM at 84 days ($p<0.05$). However, the decline was not continuous and there was an anomalous minimum in the profile at 4-6 hours. The concentration of BMP-7 in the buffer quickly increased to 9.7±2.2 ng/g DBM at 2 hours, remained nearly constant to 24 hours, then increased in an approximately linear fashion to 146±36.7 ng/g DBM at 84 days, indicating that elution was not yet complete at this interval. Thus, there was an apparent net increase in the amount of BMP-7 contained in the system of 135.2±36.8 ng/g DBM ($p<0.001$).The results, while unusual, corroborate previously published short-term (7 day) data. A two-compartment model of the distribution of BMPs in DBM, as well as a critical examination of the guanidine-HCl extraction method, is proposed to address the unique characteristics of the elution profiles.

Chapter VIII - Treatment of many oral and maxillofacial lesions necessitates intraosseous access that is obtainable only at the expense of removing the maxillary or mandibular cortical, alveolar, or basal bone. Bone removal in the oral and maxillofacial region, especially in large, expansile, maxillofacial lesions, may lead to clot breakdown, facial deformity, inadequate bone formation, and delayed or insufficient healing of the defect, necessitating future reconstructive procedures with grafts. In edentulous patients, osseous surgery may lead to loss of alveolar bone height or width, secondary surgery then being required to treat the resultant postoperative sequelae.

The composite osteoplastic flap, consisting of hard and soft tissues, is sometimes feasible with the intention of preserving the overlying bone thus, obviating the aforementioned sequelae. In this article, the applications of this flap in selected oral and maxillofacial procedures are evaluated and the advantages and clinical results are described.

In: Trabecular and Cortical Bone
Editor: V. Subramanium, M. Bhatnagar
ISBN: 978-1-62808-766-6

Chapter I

Trabecular Bone: Morphology, Biomechanics and Clinical Implications

Ebrahim Bani Hassan ***and Ali Ghasem-Zadeh***
Faculty of Veterinary Science, The University of Melbourne, Australia
Department of Medicine, Austin Health,
University of Melbourne, Australia

Abstract

The mammalian skeleton is mainly composed of *a.* cortical (compact) bone (CtB), and *b.* trabecular bone (TbB) also known as cancellous and spongy bone. The distribution of these two has evolved to provide ultimate bone strength, flexibility or stiffness where most needed. TbB is usually found in flat bones, cubic/short bones (e.g. of skull, pelvis, vertebrae, …) and the extremities of long bones. There is no definitive border between CtB, subchondral bone (ScB; a variant of CtB) and TbB and their borders usually change - regress and progress - in association with aging, loading, disease, growth or nutritional status. As we will explain, this interface is biologically the most active location in the bone, where most of bone modelling and remodelling processes are going on.

* E-mail: bani@unimelb.edu.au.

Anatomically TbB is in close contact and interaction with bone marrow – which lies in the inter-trabecular spaces. Due to its vast surface area, TbB also provides the most accessible source of minerals mid- to long term. General principles of bone resorption and formation also apply to TbB. However remodelling in TbB is predominantly hemi-osteonal (surface erosions or trenches), and hence bone surface area dependent - unlike CtB which is remodelled in osteonal (cylinder) format. Having less apparent density makes TbB lighter and less stiff, but not essentially less tough. The biomechanical properties of TbB are location-dependent and vary from flexible and tough (e.g. ribs) to fairly stiff (e.g. vertebral bodies). These properties are determined partly by the amount (percentage) of CtB and ScB in a particular bone. Biomechanically TbB is fatigue resistant - i.e. resists crack formation and dissemination due to repetitive loading, a feature that partly differentiates it from CtB and ScB - though some fatigue occurs in TbB. On the other hand, TbB is not fracture resistant. Especially in osteoporotic states (where trabeculae are partially or mostly resorbed and the support of the CtB is minimal due to thinning of cortex) fractures do occur commonly. The most well-known examples of these fractures are femoral neck and vertebral body pathologic fractures in the elderly. This chapter is aimed at providing the basic understanding of physio-anatomy and functional structure of TbB and its interaction with ScB, CtB and soft tissues. Also the mechanisms by which pathology initiates and progresses in TbB, with an emphasis on osteoporosis are further discussed.

Introduction

At the macroscopic level there are two types of bones: cortical and trabecular; however, subchondral bone is usually considered a continuation and variant of cortical bone that lies beneath articular cartilage (Figures 1& 2). Although trabecular bone comprises only about 20% of the bone mass, it represents around 80% of the bone surface. Different bones and sites within skeleton have different ratios of CtB and TbB. The vertebrae are mainly composed of trabecular bone (CtB/TbB ratio 25/75), but the ratio is 50/50 in femoral head and 95/5 in the radial diaphysis. TbB is mainly found at the metaphysis and epiphysis of long bones and within the cortices of flat bones or vertebrae [1-3]. Bone is remodelled (renewed) by osteoclasts (resorbing cells) and osteoblasts (forming cells) on the bone surface, combined as basic multicellular units (BMUs). The trabeculae play an important role by providing vast bone surface to initiate bone remodelling. As age advances, especially after menopause, remodelling becomes imbalanced and accelerated;

larger numbers of remodelling units, each depositing a smaller volume of bone than they remove (negative modelling), cause structural deterioration [4, 5]. Understanding the structure and functions of trabecular bone is essential to comprehend the processes associated with various disease conditions [6-10].

1. Trabecular Bone Anatomy and Morphology

1.1. Trabecular Bone Embryology and Growth

During fetal development and growth, long bones are generally formed as cartilaginous models surrounded by a perichondral membrane. Then the cartilage model is mineralised and converted first at its centre (middle) and later at epiphyses into the primary and secondary ossification centres, respectively (Figure 1). The perichondral membrane converts into periosteum along diaphysis which incorporates to the primary ossification centre formation and builds up the cortex during the growth period. The ossification centres gradually encroach on the remainder of the cartilage model and replace it with bone except at the growth plates, which continue to build up bone until maturity. This process is called endochondral ossification (Figure 1) [11-13].

At the metaphysis, chondrocytes at each side of the growth plate are arranged as columns and each column contains distinct zones called resting, proliferating, pre-hypertrophic and hypertrophic zones (in order of distance from the middle of the growth plate). The columns of chondrocytes and their extracellular matrix start to mineralize at the hypertrophic zone, and then hypertrophic chondrocytes die. At this stage blood vessels appear in high density in the area and provide the area with remodelling requirements including osteoblast and osteoclasts. The osteoclasts remove the remaining cartilage matrix and cell debris in between the longitudinal mineralised matrices. Then osteoblasts use the remaining mineralised matrices as a scaffold to build up layers of bone (lamellae) and make up early trabecular-like bone. The process is highly dependent on growth hormone and leads to longitudinal bone growth. Then this structure is further remodelled (resorbed and rebuilt) to construct mature trabecular bone with differently organised collagen fibres and trabeculae oriented towards the main axis of load. This continues throughout the growth period until maturation, in which secondary centres of ossification are ossified and longitudinal bone growth is arrested permanently [12, 14-26].

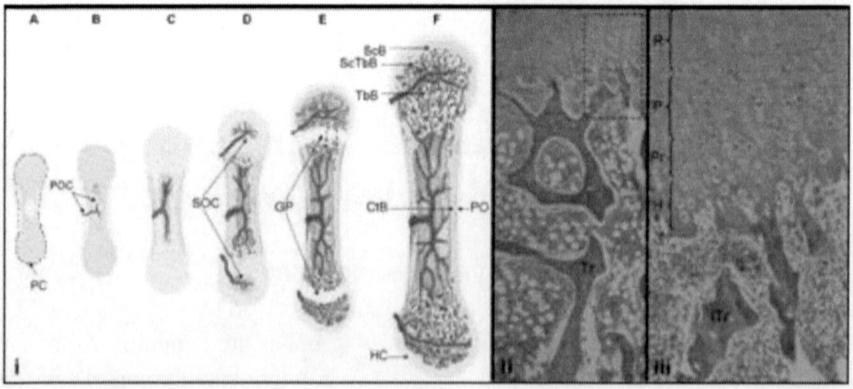

Figure 1. Endochondral ossification in schematic (i) and histologic (ii & iii, from a growing dog's growth plate) views. i:A-D occur during fetal development. i:E belongs to the growth period where growth plates are still open and i:F is a mature bone with mineralised growth plates. ii represents the region of interest outlined on the growth plate of i:E. iii is a higher magnification of the region outlined in ii showing the four stages of chondrocyte development and conversion to trabecular bone. PC: perichondrium, POC: primary center of ossification, SOC: secondary centres of ossification, GP: growth plate, HC: hyaline cartilage, ScB: subchondral bone, ScTbB: subchondral trabecular bone, TbB: Trabecular bone, CtB: cortical bone, PO: periosteum, Tr: trabecula, iTr: immature trabecula containing remains of growth cartilage, that will be replaced by fully mineralised trabecula(e), R, P, Pr and H represent: resting, proliferating, pre-hypertrophic and hypertrophic phases of endochondral osteogenesis (ii & iii histology slides courtesy of Dr Barbara Bacci).

Another process of bone formation (that produces both CtB and TbB) is by intramembranous ossification that occurs mainly in the bones of the skull where mesenchymal cells directly convert into osteoblasts. Then osteoblasts form trabecular bone centrally and cortical bone peripherally [13].

Woven bone which is mainly deposited in fracture repair, and depending on the location is usually converted into trabecular or cortical bone by remodelling. Trabecular bone has evolved to deliver a light structure with maximum strength, which supports denser CtB and ScB internally (see biomechanics of TbB below). Due to its mechanically delicate structure, TbB is not found superficially and is located axial/interior to long/flat bones with trabeculae aligned along the main axes of load [3, 27] (Figure 3).

1.2. Trabecular Bone Structure

Trabecular bone is mainly found in the following areas: I. metaphyses and epiphyses of long bones and axially within the diaphyses of some long bones (Figure 2), II. as filling inside short and flat bones III. tendon attachment protuberances [3, 27]. Despite consisting of the same constituents, TbB has different mechanical properties compared to compact bone; e.g. TbB is less stiff and has lower Young's modulus [27] which potentially can make it tough[1] in mechanical terms. Trabeculae are built and maintained as plates and rods averaging 50 to 400 μm in thickness (Figure 4) [1, 28].

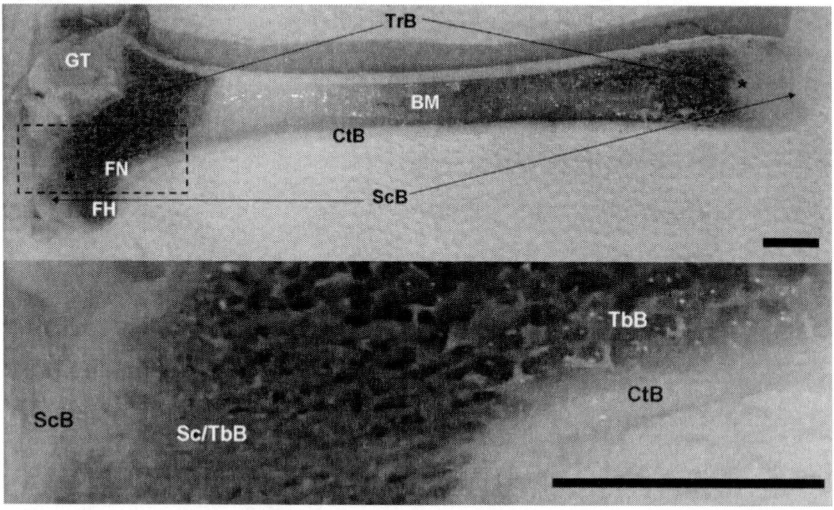

Figure 2. Upper panel: A macroscopic image of an oblique frontal section of a canine femur. Lower panel: Digital magnification of the region of interest shown above that provides a closer view of subchondral (ScB), trabecular (TbB), cortical (CtB) and subchondral trabecular (Sc/TbB) bones. Note the gradual conversion of dense ScB into TbB through Sc/TbB. BM: bone marrow, GT: greater trochanter, FN: femoral neck, FH: femoral head. Stars indicate Sc/TbB in proximal and distal femur.

Strength of trabecular bone is mainly determined by number and thickness of trabeculae and their connections [29]. Trabecular architecture (see Trabecular bone morphometry below) has been considered as a determinant

[1] In mechanical terms stiffness is different from toughness. Stiffness defines the rigidity of an object. Stiff materials resist changes in shape (deformation) in response to a load (force). On the other hand, toughness is resistance against crack development and fracture/rupture. For example, a granite stone is a stiff material and a piece of rubber is a tough one with minimal stiffness.

factor in osteoporosis and other pathological conditions of bone. Plates comprise the majority in young individuals' loaded trabecular bones, and rod model prevails in older osteoporotic cases [30-32].

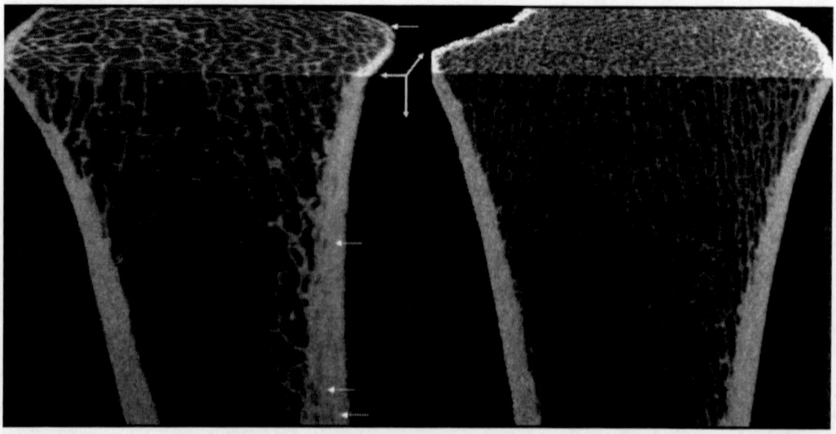

Figure 3. High Resolution-peripheral QCT (HR-pQCT; 82 μm resolution) 3D image from the tibia of two apparently healthy males *in vivo* (left: 47 and right: 32 years old). Decreased number of trabeculae and average thickness and increased trabecular separation resulted in decreased bone density of the older patient and put him at marginal status of osteoporosis/osteopenia. Despite a thicker cortex, quantitative analysis showed that the cortex had a lower grey scale value, and in closer view, trabecularization of the cortical bone (arrows) could be seen which was converting the cortical bone to "trabecular-like" bone.

Organisation of voids in bone is very sophisticated and strictly managed. Trabecular surfaces are kept smooth (and free of defects like indentations or notches) to avoid stress raisers. This is carried out by osteoblasts, which, after smoothening eroded bone surface, enter a dormant phase and convert into bone lining cells.

There is a delicate equilibrium between TbB, ScB and CtB, and the border of TbB with the other two types is surprisingly dynamic. Hormonal factors, loading and aging usually imbalance this equilibrium towards either further formation or resorption. With increased formation on trabeculae and increasing trabecular thickness and unification of multiple trabeculae (see "Trabecular bone interaction with cortical and subchondral bone" below) bone sclerosis occurs - usually at the ScB border and in response to loading or conditions such as osteoarthritis [33-38]. In contrast, progression of TbB into CtB and ScB occurs due to resorption on trabecular surfaces; and also by burrowing cortical bone, and consequently converting it into trabecular-like

CtB (trabecularization) and finally TbB (Figure 3). This mainly occurs due to aging and/or unloading and in some cases due to hormonal imbalances (e.g. sex hormone deficiency and hyperparathyroidism). Regression of cortical bone due to partial resorption will lead to decreased cortical thickness and increased cortical porosity, followed by decreased bone strength [4, 39, 40].

Figure 4. Left: 3D image of a tibia of a young dog scanned using MicroCT (10.5 μm resolution) *ex vivo*. Right: The closer view of the region outlined on the image on the left. This region contains both rod and plate models of trabecular bone. White arrows point at the plates and yellow arrowheads show rods of TbB.

1.3. Trabecular Bone Morphometry

Trabecular microarchitecture is an important determinant of bone quality and a useful indicator of fracture risk assessment due to pathological conditions in bone, especially osteoporosis [41, 42]. Trabecular thickness (TbTh) and separation (TbSp), and number of trabeculae per millimetre (trabecular number, TbN) are simple metric variables which provide further information about trabecular dimensions (Table 1 and Figure 4).

Bone surface is a variable which indirectly provides information about structure of trabeculae, and also potentially available surface for bone remodelling - as bone surface usually increases with trabecularization of CtB and ScB. The structure model index (SMI) has been formulated to provide an objective non-metric understanding of the 3D shape of the trabeculae. SMI of ideal rods and plates are 3 and 0, respectively and negative values represent plates with concave surfaces [43, 44]. The degree of anisotropy (DA) has been defined to determine how highly oriented trabeculae are within TbB. The connectivity density (ConnD) variable has also been designed to provide information on the density of connections between the trabeculae, and the idea is that with thicker trabeculae, more connections are expected between them; but resorption as small pockets, can also increase the TbN and ConnD by perforation of plates. This makes these variables less useful per se, and without considering other architectural variables (Table 1) [22, 27, 44,45].

Table 1. Definition and description of common trabecular bone microarchitecture variables

Abbreviation	Variable	Description	Standard unit
BAD	Bone apparent density	Average density of all mineralised and soft tissues within a given volume of bone	g/cm^3
BMD	Bone mineral density	Density of mineralised tissue within a given volume of bone	g/cm^3
TV	Total volume	Volume of the entire region of interest	mm^3
BV	Bone volume	Volume of the region segmented† as bone	mm^3
BS	Bone surface	Surface of the region segmented† as bone	mm^2
BV/TV	Bone volume fraction	Ratio of the segmented† bone volume to the total volume	%
TbN	Trabecular number	The average number of trabeculae (per unit of length)	1/mm
TbTh	Trabecular thickness	Mean thickness of trabeculae	mm
TbSp	Trabecular separation	Mean distance between trabeculae	mm
BS	Bone surface	Surface of the region segmented† as bone	mm^2
ConnD	Connectivity density	Degree of connectivity of trabeculae (normalised by TV)	1/mm
SMI	Structure model index	An indicator of the structure of trabeculae; SMI will be 0 for parallel plates and 3 for cylindrical rods	§
TbSpSD	Standard deviation of trabecular separation	Measure of the homogeneity of trabecular separation, assessed using direct 3D methods	§

†Segmentation of a digital image is the process of partitioning it into multiple segments based on their grey value (sets of similar pixels between certain upper and lower thresholds). The intention is to simplify the representation of an image by locating the similar pixels. This will assist visual image analysis. §Dimensionless variable.

2. Pathophysiology of Trabecular Bone

Bone formation and resorption lead to changes in the bone volume fraction (BV/TV) and in the trabecular properties. Resorption and formation are steadily carried out in bone as homeostatic and adaptational processes. Two concepts “modelling” and “remodelling” have been introduced to make the net trend of changes in bone easier to understand. Strict mechanisms control modelling and remodelling rate in bone which are affected by age, nutrition, sex, hormonal statuses, availability of mineral resources and growth etc.

Histologically, modelling refers to non-coupled formation or resorption of bone in different anatomic sites resulting in a change in the external contour of a bone or its elements i.e. cortices and trabeculae [27, 46-49]. Remodelling is defined as coupled resorption and formation (replacement) of bone through basic multicellular units (BMUs) which are discrete temporary histologic structures [27, 46, 49-51]. Riggs (2006) simply defines modelling as “removal or addition of bone to an existing surface” and remodelling as “removal of bone from one location and replacement at the same location” [46].

Remodelling in cortical bone occurs in osteonal (cylinder) fashion, but in trabecular bone it is predominantly through surface erosions or trenches (hemi-cylinders) also known as hemi-osteonal remodelling [21, 52]. Concentration of remodelling activity at a site of damage will result in bone repair, often termed targeted remodelling [53]. Trabecular bone provides ample surface for resorption and has a high blood supply, which aids in deposition and recruitment of minerals and makes it an important target organ for many calcium homeostasis hormones [3, 15, 54]. In fact, it has been shown that in aging women due to negative modelling and complete resorption of trabeculae, TbB surface decreases. Subsequently resorption and trabecularization of CtB compensates this process and creates more trabecular bone, which creates more surface area available for further resorption at the expense of thinning cortex [55].

Aging, Nutrition, Sex, Hormonal Statuses and Mineral Resources Affect TbB

Bone volume fraction and mineral density at maturation depend on multiple factors. Genetics, nutritional status of growing human/animal and availability of nutrients - minerals in particular - play a crucial role in this process. Macrominerals such as calcium and phosphorus (as building blocks of mineralised tissue), and trace minerals (such as copper, zinc, chromium which play several roles mainly as components of hormones and enzymes) are involved in bone growth [15, 56]. After maturation and under the effect of the sex hormones during the prime period (e.g. between 25-35 yrs for humans), bone reaches a life-long maximal level of density known as "peak bone mass" [9, 57-59]. The peak bone mass also correlates with the abovementioned factors which determine bone density during growth [60, 61]. After the prime period and with increasing age, bone density diminishes, and again the rate of bone loss depends on the factors which affect bone formation during growth and maturation [9, 57-59]. A person with lower peak bone mass and thinner trabeculae and cortex, compared to a person with higher bone mass would proportionally lose more bone at similar resorption rates. This would mean earlier initiation of age-associated osteoporosis in those with lower peak bone mass.

With aging and resorption of thin trabeculae followed by conversion of CtB into TbB (at endosteal side), more CtB is deposited on the periosteal surface which creates a bigger bone cross-section and better mechanical properties for the structure as a whole (Figure 5). Smaller bone size of women and also lower periosteal apposition rate compared to men, can deplete the bone of women comparatively faster than men. This occurs due to trabecularization of a higher ratio of CtB in women, rendering a more porotic bone with lower cross-sectional area. There is evidence that resorption is of similar rate in both sexes, and only architectural changes due to different resorption patterns are responsible for the differences between sexes (Figure 5) [62-64].

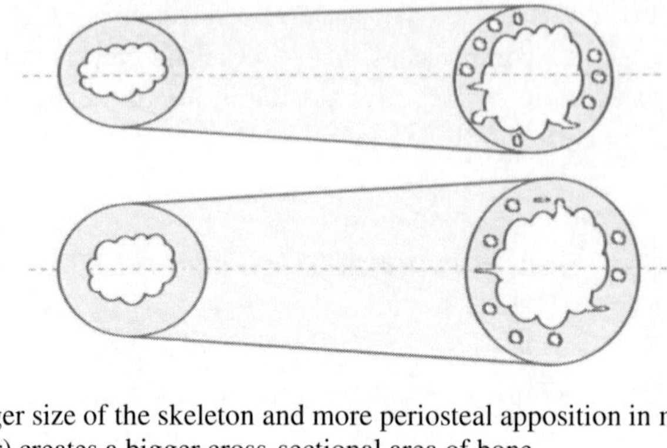

Figure 5. Bigger size of the skeleton and more periosteal apposition in men (lower) vs. women (upper) creates a bigger cross-sectional area of bone.

3. Trabecular Bone Interaction with Cortical and Subchondral Bone

The Importance of the CtB- and ScB-TbB Interfaces

The interface between TbB and the other bone types (CtB and ScB) is very dynamic. This interface changes (progresses and regresses) in response to net bone formation and resorption trend. Resorption (e.g. due to aging, unloading, loss of sex hormones, hyperparathyroidism etc.) converts dense ScB and CtB into trabecular-like bone (cortical remnant), and the process is called trabecularization. On the other hand, intensive loading or sex hormone replacement therapies can lead to thickening of trabeculae and trabecular bone sclerosis (Figure 6 & Figure 8).

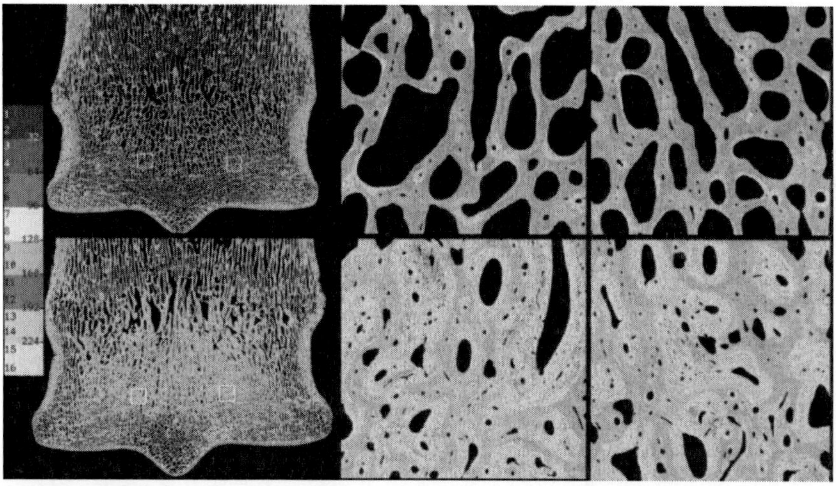

Figure 6. Quantitative back-scattered electron microscopy images of distal metacarpi of a control resting (upper panel) and a horse training only for 11 weeks (lower panel). Note densification of subchondral trabecular bone and TbB due to loading. The middle and right panels are higher magnifications of the ROIs shown on the left panel. As can be seen the newly formed layers of bone (surrounding the voids in lower image) are still of lower mineral density according to the scale on the left of the image. In severe cases and also ScB injuries, this change is extreme and TbB converts into solid bone (Figure 8). The opposite also occurs very fast and a few weeks of resting after an intensive loading period, and may lead to focal or general bone loss especially in microdamaged areas. Modified from [36] with permission.

In various animal models it has been shown that this results in progression of the interface into TbB in favour of dense CtB or ScB, which is termed corticalization [37, 55, 65-85]. This is an interesting phenomenon, as in continuous loading of ScB or some conditions such as progressed osteoarthritis or similar conditions of human and animal models, sclerosis of ScB is the most prominent feature (Figure 6 & Figure 8) [37, 86, 87].

In the case of osteoarthritis, interestingly, increased remodelling and bone loss at the early-stage, and later on slow remodelling and ScB sclerosis (modelling), are important components of pathogenicity [88]. Additionally decreased CtB thickness due to trabecularization does occur in osteoporotic conditions both focally [89] and globally [54, 55], and this can lead to lost mechanical properties and osteoporotic fractures.

4. Non-Invasive Trabecular Bone Imaging

Aging, disease, and therapeutic agents can influence the number of trabeculae and their dimensions and connectivity. Traditionally, trabecular morphometry has been assessed only in histomorphometry analysis (next topic); which is an accurate method, however it is an aggressive and destructive procedure and is limited mainly to two-dimensional analysis *ex vivo*.

On the other hand, non-invasive bone imaging and quantitative modalities such as dual energy X-ray absortiometry (DXA), quantitative computed tomography (QCT), high resolution (HR-)QCT and HR-peripheral QCT (HR-pQCT) are non-destructive methods which provide 2D and 3D images. Therefore these modalities have found frequent applications in basic as well as pre-clinical bone research throughout the last decade. These imaging modalities provide good tools for assessing compositional, morphological and functional parameters of trabecular compartments, and are playing the main role in illustrating trabecular structure in human or animals' bones. Determination of volumetric density of trabecular compartment is available by QCT and pQCT methods with ultimate resolution of 300 and 200 μm, respectively.

However imaging of trabeculae individually which span only 40-400 μm [90], requires modalities with higher resolution - such as HR-pQCT (resolution <100 μm; Figure 3), MicroCT or synchrotron radiation imaging (resolution ≤ 10 μm; Figure 4).

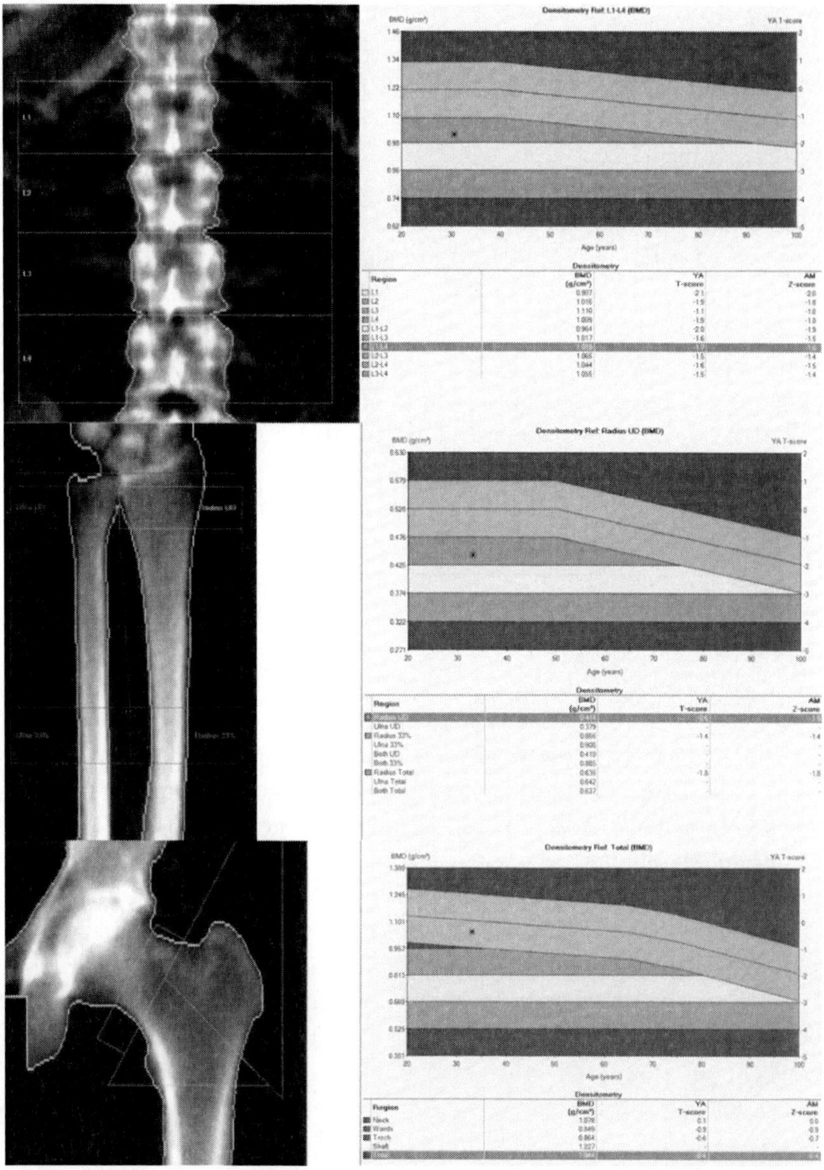

Figure 7. Dual energy X-ray absorptiometry (DXA) images of vertebral column (top) forearm (middle) and hip joint (bottom; all left panels) and their respective quantitative results and T-score graphs (right panels) in a 33 year-old healthy man. DXA is clinical standard method for assessment of the bone fracture risk by measuring bone mass on fracture prone areas. As can be seen DXA is not able to accurately differentiate trabecular bone from cortex.

Dual X-ray absorptiometry (DXA) is considered as the current clinical standard method for assessment of the bone fracture risk by measuring bone mass on fracture prone areas i.e. the spine, hip and forearm (Figure 7). This is carried out by comparing the bone mineral density of the person to sex, age and ethnicity matched controls.

Conventional DXA is incapable of differentiating CtB from TbB (Figure 7), but recently a new method has been developed to quantify the trabecular pattern as a fracture risk index. Trabecular Bone Score (TBS) is a textural parameter that can be applied to DXA, which quantifies the local variations in gray level obtained from the gray scale DXA images [91].

The T- or Z-Scores (i.e. how many standard deviations an individual's bone density is above or below the mean of a reference population) is used to calculate the fracture risk. Reference population is a large number of people with a similar ethnic background and gender.

When the reference group's average age is the same as the patient the variable is called Z-score. But when the patient is compared to a younger reference population (which are at their prime bone density), the variable is called T-score. Interpretations of DXA results have been covered somewhere else in this book.

The ability of bone to resist fracture is known as bone strength, and bone mineral density and architecture are the best predictors of fracture risk. However, bone strength can only be partially explained by bone mineral measurements by DXA [28], and therefore quantitative evaluation of trabecular structural characteristics will improve our ability to understand the physiopathology of bone disorders and in particular osteoporosis.

At the present HR-pQCT is the only method that provides images with spatial resolution in the range of the trabecular scale for human *in vivo* studies (Figure 3). HR-pQCT with 82 μm isotropic voxel size (human TbTh: 40-400 μm) is an advanced imaging technique, which provides new insights into trabecular and cortical bone structure. HR-pQCT is capable of assessing large regions of interest of distal radius, tibia or other bones, and may help indirectly to assess bone structural quality - a key component of bone strength.

Low radiation exposure (<5 μSv), quite short acquisition time (2.8 minutes), and producing 110 slices with 82 micron slice thickness which provide a 3D image of the scanned area, have made it a well suited modality for bone micro-architectural analyses *in vivo*.

HR-pQCT provides total, cortical and trabecular true volumetric density and parametric or non-parametric trabecular bone structure indices (Table 1) [92-94]. Pialat et al. have develped an analysis method which provides local topological rod- and plate-like trabeculae and may be proved helpful to improve fracture risk prediction [95]. Other imaging modalities such as synchrotron radiation imaging, electron microscopy, or Fourier transformed infrared spectroscopy (FTIR) are under investigation and presently are only available for research on small pieces of bone *ex vivo*. MRI and MicroMRI are also emerging as modalities to depict trabecular bone details in micro-scale but their applications are limited to research use at the moment [96, 97].

5. Histology of Trabecular Bone

Histology despite its aggressive nature (which requires bone biopsy or post-mortem sampling) has provided the basis for the study of the bone. Most basic variables such as trabecular properties (TbTh, TbSp and TbN) were first defined using 2D histologic images. Later on such variables were extrapolated and standardized for 3D, using montaged images of thin serial bone sections, or development of 3D imaging systems [98].

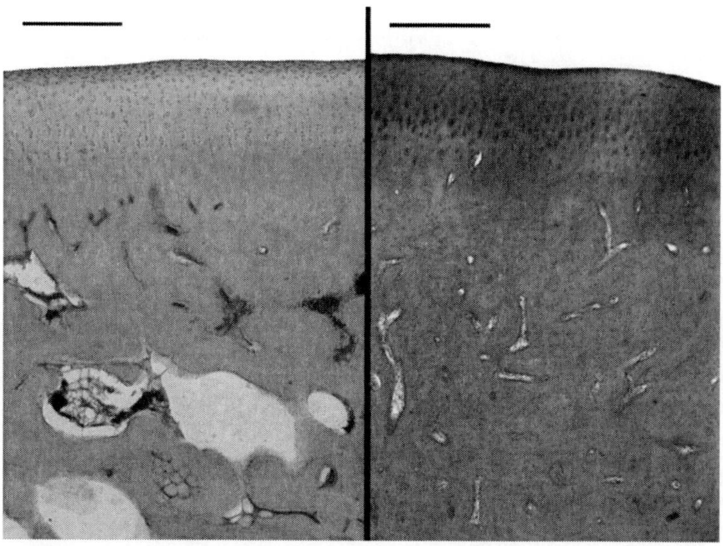

Figure 8. Comparison of the 15 μm thick cryosections of an unloaded ScB and underlying trabecular bone (left) to a similar area of an athlete horse with minor ScB pathology (right) stained with Haematoxylin and Eosin (H&E). Due to thickening and finally joining of the trabeculae, the subchondral TbB becomes sclerotic and converts into cortical-like bone with voids similar to the Haversian system of CtB (bars=500μm).

Bone histology is used for determination of trabecular architecture and more importantly bone cellular density/activity. At the moment the only direct way of visualising bone cells is using different methods of microscopy and the most common techniques are classic histology and cryohistology.

The only validated method for microfracture density determination is *en bloc* staining of microfractures to avoid inclusion of iatrogenic micro/fractures [99]. With the existing technology, quantification of osteoid surface (on Masson-Goldner stained slides), osteoblast, osteoclast and live osteocyte density and detection of bone non-mineral constituents are solely possible using histopathology techniques including histochemistry and immuno/histochemistry. Histological sections of demineralised (Figure 8) or undemineralised bone are used to quantify osteocyte injury and loss, vascular reaction, resorption, microfracture accumulation and repair response [25, 100-104]. Histology uses sections of bone as thin as 0.1 μm (transmission electron microscopy) to a few millimetres thick (undemineralised sections of bone).

Of special interest to bone biologists are the scanning and backscattered scanning electron microscopy technologies, which have been used for detection of microinjury, formation and resorption on bone samples. The artefactual microcracks due to processing of the specimen are a major problem with the technique. Therefore usually hypermineralised microcracks [36, 105, 106] or those intimately associated with resorption spaces can be considered to be original and non-artefactual.

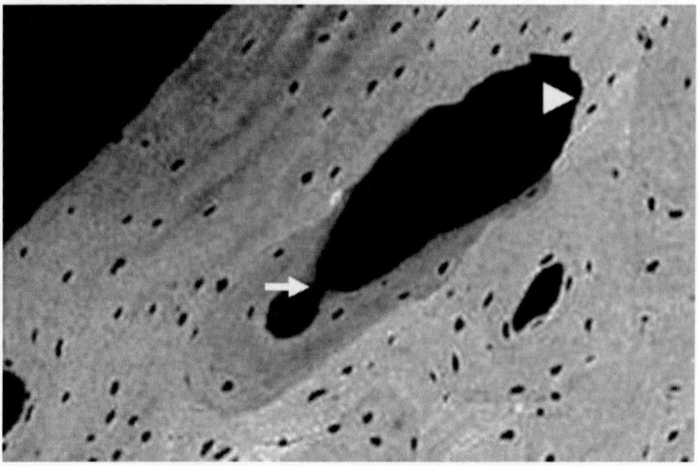

Figure 9. A backscattered scanning electron microscopy (BSEM) image of bone displaying scalloped (eroded) resorption surface (arrow head) and low density formation surface (arrow). Modified from [36] with permission.

On backscattered scanning electron microscopy images, bone with low grey value (low in calcium density and darker) parallel to the surface lamellae is considered as mineralizing surfaces and those scalloped surfaces with interrupted lamellae are considered as eroded surfaces [36, 107, 108] (Figure 9). Such surfaces are abundant on an actively remodelling TbB.

6. Trabecular Bone Endocrinology and Interactions with Soft Tissue

Due to its abundant blood supply and available surface for formation and resorption [3], it appears that TbB is metabolically most active and interactive bone structure when compared to CtB and ScB. The complexes of bone cells (osteocytes, bone lining cells, osteoblasts and osteoclasts, which the latter three are much more abundant in TbB) are coordinated and controlled by complex neuroendocrine and paracrine systems.

Due to vaster surface, greater blood supply and higher cell density, TbB is considered as the main target and reactive organ for most hormones and paracrine factors. So, as far as endocrinology of bone is concerned TbB possibly plays the most important roles. Also the TbB and compact bone are easily convertible to each other and most principles of tissue interactions and endocrinology may apply to both. Another important characteristic of TbB is an intimate proximity with bone marrow and adipose tissue, which may facilitate the crosstalk between them. Bone endocrinology and interaction with soft tissue has been covered elsewhere in this book, therefore the reader is referred to the respective chapters.

7. Trabecular Bone Biomechanics

Biomechanics is a field of science that uses physical and mathematical principles to better understand living structures and their functions.

Probably the most tangible examples of structures that provide light weight and strong mechanical properties are electric towers. Trabecular bone provides a similar structure. Now imagine ensheathing such a load resistant structure (TbB) with a cylinder of compact material (CtB). This significantly increases the structure's overall strength while giving it a light weight, and also provides further protection for the individually thin and easily breakable trabeculae [27, 109, 110].

As mentioned above, TbB has evolved to provide the maximum strength with minimal weight. Evolution has cleverly fulfilled this requirement with minimal build-up and maintenance expenses. This has been achieved by depositing the environmentally most abundant, crystallizing, insoluble-in-neutral-pH ions, i.e. calcium, phosphate and carbonate (as hydroxyapatite crystals; HA) on an organic matrix. Other than its unique mechanical properties, HA can easily be dissolved and redeposited by only changing local pH and some enzymatic activity. HA on a collagen matrix provides “material strength” for the skeleton, but to achieve “structural strength”, trabeculae are arranged parallel to the principle strain directions. These strains are imposed by daily loadings and the trabeculae are dynamically re-aligned with changing loading angle in a bone, which provides more resistance to the new loading conditions (Wolff's law) [18, 111, 112].

The evolutionary trend, i.e. change in direction of trabeculae can also be followed from apes through *australopithecines* (human like apes) to *Homo sapiens* (modern human). For example with evolution and appearance of more developed *homo* species which had erected posture the trabecular direction has evolved accordingly [113, 114]. This has given the trabeculae and consequently the bone a maximal strength, by organizing them perpendicular to the main direction of loads.

Loading and TbB

Studying biomechanics of TbB independent of cortical and subchondral protective layers is hard to justify and conduct. There are close anatomical, mechanical and biological associations between TbB, CtB and ScB, and anatomically one kind converts to the other gradually, taking intermediate shapes (Figure 2 & Figure 6). Also cutting blocks of pure TbB or test-loading of such blocks leads to buckling of superficial trabeculae, and interferes with distribution of forces associated with loading. Consequently, mechanical behaviour of pure TbB is to a great extent unknown. TbB response to loading, although it largely depends on the surrounding ScB and CtB, it displays a different response to loading and fracture repair compared to dense bone. These properties are ascribed to the architecture (structural level) rather than the mineral content (material level) of the TbB [8, 115-117].

TbB closest to loaded areas (e.g. subchondral TbB), is the interest of most investigations to study the role of microfractures in TbB properties [118-124]. Intensive studies of TbB microfractures have been done on proximal femur and vertebral TbB, due to abundance of fragility fractures and osteonecrosis occurring at those sites [10, 123, 125-127]. Reportedly microcrack density in femoral head TbB increases in elderly women with age; and those with femoral fractures had twice as many microfractures as controls. This was accompanied by decreasing bone volume fraction with increasing age [128], which rendered bone even more fragile.

Microdamage in human vertebral trabecular bone occurs as linear, diffuse and crosshatched microcracks and possibly within trabecular packets of remodelled bone rather than older non-remodelled areas [129-132]. In bovine a tibial trabecular bone model fatigue-loaded *in vitro*, diffuse damage initiates before any yield [132]. The Microfracture counts and diffuse damage accumulate and only reach significant amounts at the yield threshold, irrespective of stress magnitude [131, 133].

There are reports on the role of fatigue in trabecular bone fractures [134, 135] and abundance of microcallus in repair of such injuries [105, 117, 136, 137], but others consider TbB fatigue resistant [131]. Our observations in a horse model with fragmented ScB due to *in vivo* fatigue or in those cases with fatigue fractures of distal limb bones does not imply abundance of microfractures or microcallus formation in trabecular bone, if at all [138] (see "TbB mechanical injury and repair" below).

Application of anti-resorptive drugs increases microfracture density in trabecular bones of dog models [139, 140], consistent with the remodelling (resorption) inhibiting properties of bisphosphonates that potentially prevents the resorption of effete bone. It has been shown that antiresorption medication (bisphosphonates) increase microdamage in the vertebrae, ilium, and femoral necks of a canine model; accompanied by increased TbB volume and vertebral strength, but decreased energy absorption [141].

Finite element analysis (FEA) is a mathematical methodology for elucidating a formula for material (e.g. concrete, metal bridge, bone and/or cartilage) under a certain load and/or stress. This is done by connecting many simple element (finite elements, FE) equations of loads and stresses over many small subdomains of material, to approximate a complex equation over a larger domain. In this methodology structure is divided into "finite elements" via "nodes" in an attempt to find a relationship between external loads and internal displacements of the elements and nodes. Knowing the material properties of the material, internal strain and stress can be calculated. FE models usually are combined with experimental studies to validate the models. This technique is used to interpret and predict the final distribution of loads and stresses when a particular domain of bone is loaded. FEA can be used to predict weakness points in a given image of bone, crack and/or fracture initiation and dissemination mechanisms [109, 142, 143].

FEA nowadays is also used to simulate the mechanical behaviour of high-resolution 3D images (such as MicroCT) of bone and predict an all-inclusive equation for the given structure's behaviour. The ideal clinical application of FEA would be a more precise prediction of the probability of changes (such as fractures) in a given bone image. For example this method has been used for implant-TbB and -CtB interface stress predictions and also on clinical HR-pQCT and MicroCT scans of vertebral body and femoral neck for analysis of strength-structure relationship [142, 144, 145].

TbB Mechanical Injury and Repair

TbB fractures are of particular importance in vertebrae, where TbB comprises a high percentage of bone volume. Buckling of thinned and perforated osteoporotic trabeculae with minimal trauma, and consequent compression fractures occur commonly in vertebrae and lead to changes in stature (e.g. different degrees of kyphosis) and painful back [27, 129, 146]. In many non-traumatic fractures of bone with a trabecular compartment, the failure at least partly can be ascribed to lack of internal support by the trabecular component.

Some degree of microfracture development is reported to be the result of normal daily activity, and of common occurrence in human femoral neck and vertebral TbB [135, 147, 148]. On the other hand accumulated TbB microfractures (fatigue) reportedly can compromise the trabecular structure and lead to femoral fracture [149]. Others consider TbB fatigue-resistant [131]. Our observations in a naturally occurring large animal *in vivo* fatigue model also agrees with fatigue-resistant behaviour of TbB. Based on our observations TbB - only a few hundred micrometres away from fatigue damaged ScB - does not show any sign of microfracture accumulation or repair.

The differences in TbB fatigue resistance determination between investigators could arise from: *a*. Trabecular bones used in different studies could be of different qualities (in terms of trabecular thickness, mineral density etc.) and this potentially can affect the behaviour of tested specimens, *b*. Shock absorbing properties of ScB and cartilage in impact loading [150] might spare underlying TbB from damage in both *in vivo* and *in vitro* experiments, *c.* Higher remodelling capacity of TbB compared to ScB (due to vaster surface *in vivo*), can lead to faster repair (replacement) of such lesions; therefore the time of sampling in regards to recent loading can significantly impact the microdamage density in a given specimen.

Interesting mechanisms are involved in TbB repair, of which microcallus formation and bridging of the microfracture seem to be the most important mechanisms involved. Seaming of microcracks using hypermineralised bone has also been reported for bone [105, 137, 151], but their significance in restoring bone strength is unknown.

8. Major Clinical Conditions That Trabecular Bone Is Affected

Osteoporosis

Osteoporosis (OP) is defined as a systemic disease characterized by low bone mass and micro-architectural deterioration of bone tissue due to imbalance in formation and resorption in favour of resorption, that renders bone porous and fragile [152-154].

The skeleton of vertebrates has evolved for millions of years, and has adapted to optimal function for the particular lifespan of each species. Humans which appeared in the very recent stages of evolution are no exceptions to this principle. But in the last two hundred years, the average life expectancy of humans has almost doubled (from 40s to late 70s and 80s) [155] due to improved welfare and disease control and treatment. As a consequence of the increased lifespan the biological structures must endure for twice as long, and that's why the gradual failure of many systems and organs after 50s is a known phenomenon. For the same reason, age-associated skeletal diseases such as osteoporosis and osteoarthritis are major concerns for the aging human population.

Osteoporosis is related to the decrease of trabecular number or trabecular thickness in the bone, and to a lesser extent, to the resorption of cortical bone [90]. Due to higher bone surface and bone turnover in TbB vs. compact bone, the first signs of bone loss appear on trabecular compartment as thinning and/or complete resorption of trabeculae.

Khosla et al. [7], reported that although in aging men and women similar rate of TbB loss occurs, the structural basis for the decreased trabecular BV/TV is quite different. According to this report, bone loss in women leads to increased trabecular separation, but in men trabecular thickness decreases. This means that in aging women some trabeculae are completely resorbed, but in men trabeculae become thinner. TbN has a greater impact on the bone strength compared to TbTh, and this may explain the higher incidence of forearm fractures in women.

As mentioned above, especially in women the border between trabecular bone and dense (cortical) bone progresses in favour of trabecular bone and the existing trabeculae become thinner until are fully replaced by bone marrow or fat. Thinning of cortical bone further decreases the overall bone strength, and osteoporotic patients are notorious for developing fractures with minimal loads or traumas. As elaborated before, genetics, sex hormones, age, nutrition, physical activity and smoking condition are main determinants of OP severity in an individual.

Osteosclerosis and Ostepetrosis Syndromes

Osteosclerosis (increased bone density beyond normal) mainly occurs due to progression of ScB into TbB in conditions such as advanced osteoarthritis and also eburnation of ScB. This condition is also seen in ScB injury in animal models, namely in horse and rodent models of ScB injury [119-121, 156-163]. Some degree of ScB sclerosis and progression into TbB seems to be a normal adaptational feature in athletes, which load their bones repetitively under high strains (Figure 6) [164-167]. Osteopetrosis is an extreme generalised and hereditary form of sclerosis, in which most of TbB voids (marrow spaces) are converted into solid (corticalised) bone usually due to hormonal imbalances/genetic mutations [153]. Paget's disease is a focal form of the condition, with possible environmental factors involved, in which one or more bones in an individual become extremely mineralised and very brittle [168].

Conclusion

TbB provides an internal support for bones and also a vast dynamic surface for interaction with other tissues. Possibly the most important site of action for most hormones acting on bone is TbB. Structurally the TbB interface with dense bone is the most active site of skeleton which is under constant change. Pathogenesis of many conditions, especially osteoporosis and osteoarthritis are mainly active at this interface. The supportive role of TbB in absorption and dissipation of stress is significant and failure in the task can lead to failure of bone and fracture.

Acknowledgments

The authors are very grateful to Dr. Natali Krekeler for her precious time and input into this manuscript. Also we would like to thank Dr. Hossein Mokhtarzadeh for his comments and corrections in biomechanics section.

References

[1] Clarke, B., Normal Bone Anatomy and Physiology. *Clinical Journal of the American Society of Nephrology*, 2008. 3(Supplement 3): p. S131-S139.

[2] Watts, N.B., Clinical Utility of Biochemical Markers of Bone Remodeling. *Clinical Chemistry*, 1999. 45(8): p. 1359-1368.

[3] Barone, R., Anatomie comparée des mammifères domestiques. Cinquième ed. Vol. 1. 1999, Paris: Vigot Frères.

[4] Chavassieux, P., E. Seeman, and P.D. Delmas, Insights into material and structural basis of bone fragility from diseases associated with fractures: how determinants of the biomechanical properties of bone are compromised by disease. *Endocr Rev, 2007*. 28(2): p. 151-64.

[5] Seeman, E. and P. Delmas, Bone quality-the material and structural basis of bone strength and fragility. *The New England journal of medicine,* 2006. 354: p. 2250-61.

[6] Bjornerem, A., et al., Remodeling markers are associated with larger intracortical surface area but smaller trabecular surface area: a twin study. *Bone*, 2011. 49(6): p. 1125-30.

[7] Khosla, S., et al., Effects of sex and age on bone microstructure at the ultradistal radius: a population-based noninvasive in vivo assessment. *J Bone Miner Res,* 2006. 21(1): p. 124-31.

[8] Bayraktar, H.H., et al., Comparison of the elastic and yield properties of human femoral trabecular and cortical bone tissue. *Journal of Biomechanics,* 2004. 37(1): p. 27-35.

[9] Boonen, S., et al., Factors Associated with Cortical and Trabecular Bone Loss as Quantified by Peripheral Computed Tomography (pQCT) at the Ultradistal Radius in Aging Women. *Calcified Tissue International,* 1997. 60(2): p. 164-170.

[10] Fields, A.J. and T.M. Keaveny, Trabecular architecture and vertebral fragility in osteoporosis. *Curr Osteoporos Rep,* 2012. 10(2): p. 132-40.

[11] Dyce, M., W. Sack, and W. CJG, eds. *Textbook of Veterinary Anatomy.* 3 ed., ed. D. Lemelledo. 2002, Saunders: Philadelphia, PA. 10-25.

[12] Mackie, E., et al., Endochondral ossification: How cartilage is converted into bone in the developing skeleton. *The International Journal of Biochemistry & Cell Biology*, 2008. 40: p. 46-62.

[13] Gilbert, S., *Osteogenesis: The Development of Bones, in Developmental Biology.* 2000, Sunderland (MA): Sinauer Associates.

[14] Holopainen, J.T., et al., Changes in subchondral bone mineral density and collagen matrix organization in growing horses. *Bone,* 2008. 43(6): p. 1108-14.

[15] Swenson, M.J. and W.O. Reece, *Duke's Physiology of Domestic Animals.* 11 ed. 1993, Ithaca, NY,: Comstock Publishing Association, Cornell University Press.

[16] Klingensmith, W.C., 3rd, et al., Effect of growth hormone on metaphyseal uptake of 99mTc-methylene diphosphonate. *J Clin Endocrinol Metab,* 1981. 52(6): p. 1162-4.

[17] Rokkanen, P., S. Paatsama, and P. Rissanen, Hormone induced changes in the bones of young dogs. *Calcified Tissue Research*, 1968. 2(1): p. 95-95.

[18] Barak, M.M., D.E. Lieberman, and J.J. Hublin, A Wolff in sheep's clothing: trabecular bone adaptation in response to changes in joint loading orientation. *Bone*, 2011. 49(6): p. 1141-51.

[19] Decker, B., H. Bartels, and S. Decker, Relationships between endothelial cells, pericytes, and osteoblasts during bone formation in the sheep femur following implantation of tricalciumphosphate-ceramic. *Anat Rec,* 1995. 242(3): p. 310-20.

[20] Stern, P.H., et al., Endothelin Receptors, Second Messengers, and Actions in Bone, in *The International Conference on Progress in Bone and Mineral Research.* 1995, American Institute of Nutrition. p. 2028S-2032S.

[21] Parfitt, A.M., Osteonal and hemi-osteonal remodeling: The spatial and temporal framework for signal traffic in adult human bone. *Journal of Cellular Biochemistry*, 1994. 55(3): p. 273-286.

[22] Bouxsein, M.L., et al., Guidelines for assessment of bone microstructure in rodents using micro–computed tomography. *Journal of Bone and Mineral Research*, 2010. 25(7): p. 1468-1486.

[23] Burr, D.B. and E. Radin, Microfracture sand microcracks in subchondral bone: are they relevant to osteoarthritis? *Rheumatic Disease Clinics of North America*, 2003. 29: p. 675-685.

[24] Carano, R.A.D.a.F., Ellen H. , Angiogenesis and bone repair. *Drug Discovery Today*, 2003. 8(21): p. 980-9.

[25] Imhof, H., et al., Degenerative joint disease: cartilage or vascular disease? *Skeletal Radiology*, 1997. 26(7): p. 398-403.

[26] Wohl, G.R., D.A. Towler, and M.J. Silva, Stress fracture healing: fatigue loading of the rat ulna induces upregulation in expression of osteogenic and angiogenic genes that mimic the intramembranous portion of fracture repair. *Bone*, 2009. 44(2): p. 320-30.

[27] Currey, J., *Bones: structure and mechanics*. 2006, NJ: Princeton Universuty press.

[28] Zebaze, R., et al. *Cortical and trabecular bone: a dichotomy revisited. in ANZBMS 18th annual scientific meeting*. 2008. Melbourne.

[29] Seeman, E. and J.L. Hopper, Genetic and environmental components of the population variance in bone density. *Osteoporos Int,* 1997. 7 Suppl 3: p. S10-6.

[30] Laib, A., et al., 3D micro-computed tomography of trabecular and cortical bone architecture with application to a rat model of immobilisation osteoporosis. *Medical and Biological Engineering and Computing*, 2000. 38(3): p. 326-332.

[31] Mc Donnell, P., P.E. Mc Hugh, and D. O' Mahoney, Vertebral Osteoporosis and Trabecular Bone Quality. *Annals of Biomedical Engineering,* 2007. 35(2): p. 170-189.

[32] Campbell, G.M., H.R. Buie, and S.K. Boyd, Signs of irreversible architectural changes occur early in the development of experimental osteoporosis as assessed by in vivo micro-CT. *Osteoporosis International,* 2008. 19(10): p. 1409-1419.

[33] Anastasiou, A., et al., <i>Ex Vivo</i> Magnetic Resonance Imaging of the Distal Row of Equine Carpal Bones: Assessment of Bone Sclerosis and Cartilage Damage. *Veterinary Radiology & Ultrasound*, 2003. 44(5): p. 501-512.

[34] Cox, L.G.E., et al., Decreased bone tissue mineralization can partly explain subchondral sclerosis observed in osteoarthritis. *Bone,* 2012. 50(5): p. 1152-1161.

[35] Jaiprakash, A., et al., Phenotypic characterization of osteoarthritic osteocytes from the sclerotic zones: a possible pathological role in subchondral bone sclerosis. *Int J Biol Sci*, 2012. 8(3): p. 406-17.

[36] Boyde, A. and E. Firth, Musculoskeletal responses of 2-year-old Thoroughbred horses to early training. 8. Quantitative back-scattered electron scanning electron microscopy and confocal fluorescence microscopy of the epiphysis of the third metacarpal bone. *N Z Vet J,* 2005. 53(2): p. 123-32.

[37] Murray, R.C., et al., Subchondral bone thickness, hardness and remodelling are influenced by short-term exercise in a site-specific manner. *Journal of Orthopaedic Research*, 2001. 19(6): p. 1035-1042.

[38] Tranquille, C.A., et al., Effect of exercise on thicknesses of mature hyaline cartilage, calcified cartilage, and subchondral bone of equine tarsi. *American Journal of Veterinary Research*, 2009. 70(12): p. 1477-1483.

[39] Seeman, E., Estrogen, androgen, and the pathogenesis of bone fragility in women and men. *Curr Osteoporos Rep*, 2004. 2(3): p. 90-6.

[40] Lindsay, R. and F. Cosman, Osteoporosis, in *Harrison's Principles of Internal Medicine*, D.L. Longo, et al., Editors. 2012, McGrawHill: NY. p. 3120-35.

[41] Dempster, D.W., The Contribution of Trabecular Architecture to Cancellous Bone Quality. *Journal of Bone and Mineral Research*, 2000. 15(1): p. 20-23.

[42] Seeman, E. and P.D. Delmas, Bone Quality — The Material and Structural Basis of Bone Strength and Fragility. *New England Journal of Medicine,* 2006. 354(21): p. 2250-2261.

[43] b-cube AG Co. 3D Morphometry. 2013 14/6/2013]; Available from: http://www.b-cube.ch/index.php?option=com_content&view=article&id=22&Itemid=20.

[44] Bouxsein, M.L., et al., Guidelines for assessment of bone microstructure in rodents using micro-computed tomography. *J Bone Miner Res,* 2010. 25(7): p. 1468-86.

[45] Van Der Linden, J.C., J.A.N. Verhaar, and H. Weinans, A Three-Dimensional Simulation of Age-Related Remodeling in Trabecular Bone. *Journal of Bone and Mineral Research*, 2001. 16(4): p. 688-696.

[46] Riggs, C.M., Osteochondral injury and joint disease in the athletic horse. *Equine Veterinary Education*, 2006. 18(2): p. 100-112.

[47] Dyson, S.J., Navicular disease and other soft tissue causes of palmar foot pain, in *Diagnosis and management of lameness in the horse*, M.W. Ross and S.J. Dyson, Editors. 2003, Saunders: St. Louis - USA. p. 286-299.

[48] Segen, J., Modelling, in The dictionary of modern medicine: A sourcebook of currently used medical expressions, jargon and technical terms. 1992, Parthenon publishing group Ltd: Lancs - UK.

[49] Frost, H.M., A determinant of bone architecture. The minimum effective strain. *Clin Orthop Relat Res*, 1983(175): p. 286-92.

[50] Jilka, R.L., Biology of the basic multicellular unit and the pathophysiology of osteoporosis. *Medical and Pediatric Oncology*, 2003. 41(3): p. 182-185.

[51] Dyson, S.J., Diagnosis and management of lameness in the horse, in *Diagnosis and Management of Lameness in the Horse*, M.W. Ross and S.J. Dyson, Editors. 2003, W.B. Saunders Co.: Philadelphia. p. 320.

[52] van Oers, R.F., et al., A unified theory for osteonal and hemi-osteonal remodeling. *Bone,* 2008. 42(2): p. 250-9.

[53] Burr, D.B., Targeted and nontargeted remodeling. *Bone*, 2002. 30(1): p. 2-4.

[54] Iwamoto, J., et al., Response of cortical and cancellous bones to mild calcium deficiency in young growing female rats: a bone histomorphometry study. *Exp Anim*, 2004. 53(4): p. 347-54.

[55] Zebaze, R.M.D., et al., Intracortical remodelling and porosity in the distal radius and post-mortem femurs of women: a cross-sectional study. *The Lancet*, 2010. 375(9727): p. 1729-1736.

[56] Ilich, J.Z. and J.E. Kerstetter, Nutrition in Bone Health Revisited: A Story Beyond Calcium. *Journal of the American College of Nutrition,* 2000. 19(6): p. 715-737.

[57] Hannan, M.T., et al., Risk factors for longitudinal bone loss in elderly men and women: the Framingham Osteoporosis Study. *J Bone Miner Res,* 2000. 15(4): p. 710-20.

[58] Mazess, R.B., On Aging Bone Loss. *Clinical Orthopaedics and Related Research,* 1982. 165: p. 239-252.

[59] Burr, D.B., Muscle Strength, Bone Mass, and Age-Related Bone Loss. *Journal of Bone and Mineral Research*, 1997. 12(10): p. 1547-1551.

[60] Teegarden, D., et al., Peak bone mass in young women. *Journal of Bone and Mineral Research*, 1995. 10(5): p. 711-715.

[61] Lorentzon, M., D. Mellström, and C. Ohlsson, Age of Attainment of Peak Bone Mass Is Site Specific in Swedish Men—The GOOD Study. *Journal of Bone and Mineral Research*, 2005. 20(7): p. 1223-1227.

[62] Branch, M., et al., Is there a characteristic distal tarsal subchondral bone plate thickness pattern in horses with no history of hindlimb lameness? *Equine Vet J Suppl,* 2005. 37(5): p. 450 - 455.

[63] Seeman, E., Bone quality: the material and structural basis of bone strength. *J Bone Miner Metab*, 2008. 26: p. 1-8.

[64] Seeman, E., Structural basis of growth-related gain and age-related loss of bone strength. Rheumatology (Oxford), 2008. 47 Suppl 4: p. iv2-8.

[65] Swerdloff, R.S. and C. Wang, Androgens and aging in men. *Experimental Gerontology*, 1993. 28(4–5): p. 435-446.

[66] Rabijewski, M. and W. Zgliczynski, [Testosterone deficiency in elderly men]. *Pol Merkur Lekarski*, 2009. 27(162): p. 517-23.

[67] Saad, F. and J. Eastham, Maintaining Bone Health in Prostate Cancer Throughout the Disease Continuum. *Seminars in Oncology*, 2010. 37, Supplement 1(0): p. S30-S37.

[68] Smith, M.R., Androgen deprivation therapy for prostate cancer: new concepts and concerns. *Curr Opin Endocrinol Diabetes Obes*, 2007. 14(3): p. 247-54.

[69] Greenspan, S.L., Approach to the Prostate Cancer Patient with Bone Disease. *Journal of Clinical Endocrinology & Metabolism*, 2008. 93(1): p. 2-7.

[70] Higano, C.S., Androgen-deprivation-therapy-induced fractures in men with nonmetastatic prostate cancer: what do we really know? *Nat Clin Pract Urol,* 2008. 5(1): p. 24-34.

[71] Grossmann, M., et al., Bone and metabolic health in patients with non-metastatic prostate cancer who are receiving androgen deprivation therapy. *Med J Aust*, 2011. 194(6): p. 301-6.

[72] Hamilton, E.J., et al., Structural decay of bone microarchitecture in men with prostate cancer treated with androgen deprivation therapy. *J Clin Endocrinol Metab*, 2010. 95(12): p. 2010-0902.

[73] van Harreveld, Philip D. , et al., Clinical evaluation of the effects of immobilization followed by remobilization and exercise on the metacarpophalangeal joint in horses. *American Journal of Veterinary Research,* 2002. 63(2): p. 282-288.

[74] van Harreveld, Philip D. , et al., Effects of immobilization followed by remobilization on mineral density, histomorphometric features, and

formation of the bones of the metacarpophalangeal joint in horses. *American Journal of Veterinary Research*, 2002. 63(2): p. 276-281.
[75] Whitton, R.C., et al., Third metacarpal condylar fatigue fractures in equine athletes occur within previously modelled subchondral bone. *Bone*, 2010. 47(4): p. 826-831.
[76] McCarthy, R.N. and L.B. Jeffcott, Effects of treadmill exercise on cortical bone in the third metacarpus of young horses. *Research in Veterinary Science*, 1992. 52(1): p. 28-37.
[77] Muir, P., et al., Exercise-induced metacarpophalangeal joint adaptation in the Thoroughbred racehorse. *J Anat*, 2008. 213(6): p. 706-17.
[78] Firth, E.C., The response of bone, articular cartilage and tendon to exercise in the horse. *Journal of Anatomy*, 2006. 208(4): p. 513-526.
[79] Riggs, C.M. and A. Boyde, Effect of exercise on bone density in distal regions of the equine third metacarpal bone in 2-year-old Thoroughbreds. *Equine Veterinary Journal*, 1999. 31(S30): p. 555-560.
[80] McCarthy, R.N. and L.B. Jeffcott, Effects of treadmill exercise on cortical bone in the third metacarpus of young horses. *Res Vet Sci*, 1992. 52(1): p. 28-37.
[81] Horcajada, M.-N., et al., Influence of treadmill running on femoral bone in young orchidectomized rats. *Journal of Applied Physiology*, 1997. 83(1): p. 129-133.
[82] Horcajada-Molteni, M.N., et al., Treadmill running starting 3 months after orchidectomy restores femoral bone mass in rats. *Eur J Appl Physiol Occup Physiol,* 1999. 79(3): p. 251-9.
[83] Iwamoto, J., T. Takeda, and Y. Sato, Effect of treadmill exercise on bone mass in female rats. *Exp Anim*, 2005. 54(1): p. 1-6.
[84] Woo, S.L., et al., The effect of prolonged physical training on the properties of long bone: a study of Wolff's Law. *J Bone Joint Surg Am*, 1981. 63(5): p. 780-7.
[85] Gordon, K.R., M. Perl, and C. Levy, Structural alterations and breaking strength of mouse femora exposed to three activity regimens. *Bone,* 1989. 10(4): p. 303-12.
[86] Grynpas, M., et al., Subchondral bone in osteoarthritis. *Calcified Tissue International,* 1991. 49(1): p. 20-26.
[87] Buckland-Wright, C., Subchondral bone changes in hand and knee osteoarthritis detected by radiography. *Osteoarthritis and Cartilage*, 2004. 12, Supplement(0): p. 10-19.
[88] Burr, D.B. and M.A. Gallant, Bone remodelling in osteoarthritis. *Nat Rev Rheumatol*, 2012. 8(11): p. 665-73.

[89] Whitton, R.C., et al., Exercise-induced inhibition of remodelling is focally offset with fatigue fracture in racehorses. Osteoporos Int, 2013.

[90] Nogueira, L.P., et al., 3D histomorphometric quantification of trabecular bones by computed microtomography using synchrotron radiation. *Micron,* 2010. 41(8): p. 990-6.

[91] Hans, D., et al., Bone microarchitecture assessed by TBS predicts osteoporotic fractures independent of bone density: the Manitoba study. *J Bone Miner Res*, 2011. 26(11): p. 2762-9.

[92] Laib, A., H.J. Hauselmann, and P. Ruegsegger, In vivo high resolution 3D-QCT of the human forearm. *Technol Health Care*, 1998. 6(5-6): p. 329-37.

[93] Boutroy, S., et al., In vivo assessment of trabecular bone microarchitecture by high-resolution peripheral quantitative computed tomography. *J Clin Endocrinol Metab*, 2005. 90(12): p. 6508-15.

[94] Burghardt, A.J., et al., Reproducibility of direct quantitative measures of cortical bone microarchitecture of the distal radius and tibia by HR-pQCT. *Bone,* 2010. 47(3): p. 519-28.

[95] Pialat, J.B., et al., Local topological analysis at the distal radius by HR-pQCT: Application to in vivo bone microarchitecture and fracture assessment in the OFELY study. *Bone,* 2012. 51(3): p. 362-8.

[96] Weiger, M., M. Stampanoni, and K.P. Pruessmann, Direct depiction of bone microstructure using MRI with zero echo time. *Bone,* 2013. 54(1): p. 44-47.

[97] Kim, N., et al., Evaluation of MRI resolution affecting trabecular bone parameters: determination of acceptable resolution. *Magn Reson Med,* 2012. 67(1): p. 218-25.

[98] Parfitt, A.M., et al., Bone histomorphometry: Standardization of nomenclature, symbols, and units: Report of the asbmr histomorphometry nomenclature committee. *Journal of Bone and Mineral Research,* 1987. 2(6): p. 595-610.

[99] Burr, D.B. and T. Stafford, Validity of the bulk-staining technique to separate artifactual from in vivo bone microdamage. *Clin Orthop Relat Res,* 1990(260): p. 305-8.

[100] Drum, M.G., et al., Comparison of Gross And Histopathologic Findings with Quantitative Computed Tomographic Bone Density in the Distal Third Metacarpal Bone Of Racehorses. *Veterinary Radiology & Ultrasound,* 2007. 48(6): p. 518-527.

[101] Verborgt, O., G.J. Gibson, and M.B. Schaffler, Loss of Osteocyte Integrity in Association with Microdamage and Bone Remodeling After

Fatigue In Vivo. *Journal of Bone and Mineral Research*, 2000. 15(1): p. 60-67.

[102] Walsh, D., et al., Angiogenesis in the synovium and at the osteochondral junction in osteoarthritis. *Osteoarthritis and Cartilage*, 2007. 15(7).

[103] Oettmeier, R., et al., Quantitative study of articular cartilage and subchondral bone remodeling in the knee joint of dogs after strenuous running training. *Journal of Bone and Mineral Research*, 1992. 7(S2): p. S419-S424.

[104] Kim, C.W., et al., Response of fractured osteoporotic bone to polymethylmethacrylate after vertebroplasty: case report. *The Spine Journal,* 2004. 4(6): p. 709-712.

[105] Boyde, A., The real response of bone to exercise. *Journal of Anatomy,* 2003. 203(2): p. 173-189.

[106] Boyde, A. and E. Firth, High resolution microscopic survey of third metacarpal articular calcified cartilage and subchondral bone in the juvenile horse: possible implications in chondro-osseous disease. *Microscopy Research and Technique* 2008. 71(6): p. 477-88.

[107] Boyde, A., Scanning electron microscopy of bone. *Methods Mol Biol*, 2012. 816: p. 365-400.

[108] Fratzl, P., et al., Structure and mechanical quality of the collagen-mineral nano-composite in bone. *Journal of Materials Chemistry*, 2004. 14(14): p. 2115-2123.

[109] Oomens, C., M. Brekelmans, and F. Baaijens, *Biomechanics: Concepts and Computation*. 2010: Cambridge University Press.

[110] Knudson, D., Fundamentalsof Biomechanics. 2 ed. 2007, NY: Springer.

[111] Wolff, J., Das Gesetz der Transformation der Knochen. 1892, Berlin: Hirschwald.

[112] Ryan, T.M. and G.E. Krovitz, Trabecular bone ontogeny in the human proximal femur. *Journal of Human Evolution*, 2006. 51(6): p. 591-602.

[113] Macchiarelli, R., et al., Hip bone trabecular architecture shows uniquely distinctive locomotor behaviour in South African australopithecines. *Journal of Human Evolution*, 1999. 36(2): p. 211-232.

[114] Chappard, D., General review: Trabecular bone microarchitecture: A review. *Morphologie*, 2008. 92(299): p. 162.

[115] Choi, K., et al., The elastic moduli of human subchondral, trabecular, and cortical bone tissue and the size-dependency of cortical bone modulus. *Journal of Biomechanics*, 1990. 23(11): p. 1103-1113.

[116] Rho, J.-Y., L. Kuhn-Spearing, and P. Zioupos, Mechanical properties and the hierarchical structure of bone. *Medical Engineering & Physics,* 1998. 20(2): p. 92-102.

[117] Burr, D.B., The importance of subchondral bone in the progression of osteoarthritis. *The Journal of Rheumatology*, 2004. 70: p. 77-80.

[118] Pugh, J.W., R.M. Rose, and E.L. Radin, A Possible Mechanism of Wolff's Law: Trabecular Microfractures. *Archives Of Physiology and Biochemistry*, 1973. 81(1): p. 27-40.

[119] Radin, E.L., et al., Effects of mechanical loading on the tissues of the rabbit knee. *Journal of Orthopaedic Research*, 1984. 2(3): p. 221-234.

[120] Radin, E.L., et al., Response of joints to impact loading. 3. Relationship between trabecular microfractures and cartilage degeneration. *Journal of Biomechanics*, 1973. 6(1): p. 51-7.

[121] Radin, E.L., I.L. Paul, and M.J. Tolkoff, Subchondral bone changes in patients with early degenerative joint disease. *Arthritis & Rheumatism*, 1970. 13(4): p. 400-405.

[122] Muir, P., et al., Exercise-induced metacarpophalangeal joint adaptation in the Thoroughbred racehorse. *Journal of Anatomy*, 2008. 213: p. 706-717

[123] Chiba, K., et al., Heterogeneity of Bone Microstructure in the Femoral Head in Patients with Osteoporosis: An ex vivo HR-pQCT study. *Bone,* 2013.

[124] Firth, E., et al., Changes in diaphyseal and epiphyseal bone parameters in thoroughbred horses after withdrawal from training. *J Musculoskelet Neuronal Interact* 2007. 7(1): p. 74-76.

[125] Ollivier, M., et al., Radiographic bone texture analysis is correlated with 3D microarchitecture in the femoral head, and improves the estimation of the femoral neck fracture risk when combined with bone mineral density. *Eur J Radiol,* 2013.

[126] Tingart, M., et al., Analysis of bone matrix composition and trabecular microarchitecture of the femoral metaphysis in patients with osteonecrosis of the femoral head. *J Orthop Res,* 2009. 27(9): p. 1175-81.

[127] Li, H., C. Zhang, and B. Zeng, [Changes in bone micro-architecture and bone mineral density following experimental osteonecrosis of femoral head by local injection of ethanol in canines]. *Zhongguo Xiu Fu Chong Jian Wai Ke Za Zhi,* 2008. 22(3): p. 281-9.

[128] Mori, S., et al., Trabecular bone volume and microdamage accumulation in the femoral heads of women with and without femoral neck fractures. *Bone*, 1997. 21(6): p. 521-6.

[129] Wenzel, T.E., M.B. Schaffler, and D.P. Fyhrie, In vivo trabecular microcracks in human vertebral bone. *Bone,* 1996. 19(2): p. 89-95.

[130] Wang, L., et al., Three-dimensional morphology of microdamage in peri-screw bone: a scanning electron microscopy of methylmethacrylate cast replica. *Microsc Microanal*, 2012. 18(5): p. 1106-11.

[131] Donahue, S. and S. Galley, Microdamage in bone: Implications for fracture, repair, remodeling and adaptation. *Critical Reviews in biomedical engineering*, 2006. 34: p. 215-271.

[132] Keaveny, T.M., et al., Biomechanics of Trabecular bone. *Annu. Rev. Biomed. Eng.* , 2001. 3: p. 307-333.

[133] Wachtel, E.F. and T.M. Keaveny, Dependence of trabecular damage on mechanical strain. *J Orthop Res*, 1997. 15(5): p. 781-7.

[134] Michel, M.C., et al., Compressive fatigue behavior of bovine trabecular bone. *Journal of Biomechanics*, 1993. 26(4–5): p. 453-463.

[135] Todd, R.C., M.A.R. Freeman, and C.J. Pirie, Isolated Trabecular Fatigue Fractures in the Femoral Head. *Journal of Bone & Joint Surgery*, British Volume, 1972. 54-B(4): p. 723-728.

[136] Mori, S., [Bone fracture and the healing mechanisms. Microdamage and Microfracture]. *Clin Calcium*, 2009. 19(5): p. 699-703.

[137] Banse, X., et al., Vertebral cancellous bone turn-over: microcallus and bridges in backscatter electron microscopy. *Micron*, 2005. 36(7-8): p. 710-4.

[138] Bani Hassan, E , et al., Palmar/planar osteochondral disease is associated with increased bone volume fraction and microfracture accumulation, in *The Annual General Meeting of the Australia & New Zealand Bone & Mineral Society*. 2011: Goldcoast, Australia.

[139] Mashiba, T., et al., Effects of high-dose etidronate treatment on microdamage accumulation and biomechanical properties in beagle bone before occurrence of spontaneous fractures. *Bone,* 2001. 29(3): p. 271-8.

[140] Mashiba, T., et al., Suppressed Bone Turnover by Bisphosphonates Increases Microdamage Accumulation and Reduces Some Biomechanical Properties in Dog Rib. *Journal of Bone and Mineral Research*, 2000. 15(4): p. 613-620.

[141] Mashiba, T., et al., Effects of suppressed bone turnover by bisphosphonates on microdamage accumulation and biomechanical

properties in clinically relevant skeletal sites in beagles. *Bone,* 2001. 28(5): p. 524-31.

[142] Kluess, D., et al., *Finite Element Analysis in Orthopaedic Biomechanics, in Finite Element Analysis*, D. Moratal, Editor. 2010, Sciyo. p. 151-170.

[143] D Moratal, Finite Element Analysis. . 2010: Sciyo, http://www.intechopen.com/books/finite-element-analysis.

[144] Chen, L., Finite Element Analysis of the Stress on the Implant-Bone Interface of Dental Implants with Different Structures in Finite Element Analysis – New Trends and Developments F. Ebrahimi, Editor. 2012, Intech: Available online.

[145] Bevill, G. and T.M. Keaveny, Trabecular bone strength predictions using finite element analysis of micro-scale images at limited spatial resolution. *Bone,* 2009. 44(4): p. 579-84.

[146] Kleerekoper, M., et al., The role of three-dimensional trabecular microstructure in the pathogenesis of vertebral compression fractures. *Calcified Tissue International,* 1985. 37(6): p. 594-597.

[147] Fazzalari, N., Trabecular microfracture. *Calcified Tissue International,* 1993. 53(1): p. S143-S147.

[148] Frost, H.M., The spinal osteoporoses: Mechanisms of pathogenesis and pathophysiology. *Clinics in Endocrinology and Metabolism*, 1973. 2(2): p. 257-275.

[149] Freeman, M.A., R.C. Todd, and C.J. Pirie, The role of fatigue in the pathogenesis of senile femoral neck fractures. *J Bone Joint Surg Br,* 1974. 56-B(4): p. 698-702.

[150] Malekipour, F., et al., Shock absorbing ability of articular cartilage and subchondral bone under impact compression. *Journal of the Mechanical Behavior of Biomedical Materials*, (0).

[151] Radin, E.L., I.L. Paul, and M.J. Tolkoff, Subchondral Bone Changes in Patients with Early Degenerative Joint Disease. *Arthritis and Rheumatism,* 1970. 13(4): p. 400-5.

[152] Studdert, V.P., C.C. Gay, and D.C. Blood, Osteoporosis, in *Saunders Comprehensive Veterinary Dictionary*. 2012, Elsevier-Saunders.

[153] Dorland, in *Dorland's Medical Dictionary*. 2007, Saunderers-Elsevier.

[154] Whitaker, M., et al., Bisphosphonates for Osteoporosis — Where Do We Go from Here? *New England Journal of Medicine*, 2012. 366(22): p. 2048-2051.

[155] Oeppen, J. and J. Vaupel, Broken limits to life expectancy. *Science,* 2002. 296(5570): p. 1029.

[156] Goldring, M.B. and S.R. Goldring, Articular cartilage and subchondral bone in the pathogenesis of osteoarthritis. *Annals of the New York Academy of Sciences*, 2010. 1192(1): p. 230-237.

[157] Goldring, S. and M. Goldring, Bone and cartilage in osteoarthritis: is what's best for one good or bad for the other? *Arthritis Research & Therapy*, 2010. 12(5): p. 143.

[158] Goldring, S.R., Alterations in periarticular bone and cross talk between subchondral bone and articular cartilage in osteoarthritis. *Ther Adv Musculoskelet Dis,* 2012. 4(4): p. 249-58.

[159] Kawcak, C., et al., The role of subchondral bone in joint disease: a review. *Equine Veterinary Journal*, 2001. 33: p. 120-126.

[160] Radin, E.L. and R.M. Rose, Role of subchondral bone in the initiation and progression of cartilage damage. *Clinical Orthopaedics and Related Research,* 1986(213): p. 34-40.

[161] Norrdin, R., et al., Overload arthrosis: strain patterns in the equine metacarpal condyle. *J Musculoskelet Neuronal Interact*, 2001. 1(4): p. 357-62.

[162] Norrdin, R., et al., Subchondral bone failure in an equine model of overload arthrosis. *Bone*, 1998. 22(2): p. 133-39.

[163] Norrdin, R. and S. Stover, Subchondral bone failure in overload arthrosis: A scanning electron microscopic study in horses. *J Musculoskelet Neuronal Interact* 2006. 6(3): p. 251-257.

[164] Firth, E., Problems in quantifying bone response to exercise in horses: a review. *New Zealand Veterinary Journal*, 2004. 52(5): p. 216 - 229.

[165] Firth, E.C. and C.W. Rogers, Musculoskeletal responses of 2-year-old Thoroughbred horses to early training. 7. Bone and articular cartilage response in the carpus. *New Zealand Veterinary Journal*, 2005. 53: p. 113-122.

[166] Firth, E.C., et al., Musculoskeletal responses of 2-year-old Thoroughbred horses to early training. 6. Bone parameters in the third metacarpal and third metatarsal bones. New Zealand Veterinary Journal, 2005. 53: p. 101-112.

[167] Milz, S. and R. Putz, Quantitative morphology of the subchondral plate of the tibial plateau. *J Anat,* 1994. 185 (Pt 1): p. 103-10.

[168] Mirra, J., E. Brien, and J. Tehranzadeh, Paget's disease of bone: review with emphasis on radiologic features, part I. *Skeletal Radiology*, 1995. 24(3): p. 163-171.

In: Trabecular and Cortical Bone ISBN: 978-1-62808-766-6
Editor: V. Subramanium, M. Bhatnagar

Chapter II

Cortical Bone As an Enduring Depot of Bone-Seeking Contaminants – Evaluation of Cortical Bone Resorption Rate on the Basis of a Contaminant (^{90}Sr): Measurement in Humans

E. I. Tolstykh[1], N. B. Shagina[1], M. O. Degteva[1], L. R. Anspaugh[2] and B. A. Napier[3]

[1]Urals Research Center for Radiation Medicine, Medgorodok, Chelyabinsk, Russian Federation
[2]Radiobiology Division, Department of Radiology, University of Utah, Salt Lake City, Utah, US
[3]Pacific Northwest National Laboratory, Richland, Washington, US

Abstract

Cortical bone is a major part of the human skeleton and amounts to about 60-80% of the skeletal mass. The rate of cortical bone remodeling is low and elements incorporated in the crystal structure of cortical bone remain there for a long period of time. As a result, cortical bone becomes a "long-term storage" for different environmental contaminants. In this study accumulation of bone-seeking contaminants is considered by an example of strontium-90 (^{90}Sr) accumulation investigated in long-term studies at the Techa River (Southern Urals, Russia) and studies of global fallout from nuclear weapon tests. Accidental ingestion of significant levels of ^{90}Sr by residents of the Techa Riverside villages (about 30,000 persons) occurred in the 1950s as a result of the river contamination by liquid radioactive wastes. Strontium-90 contents in the skeleton (mainly in cortical part of the skeleton) have been measured *in vivo* in these residents over the last 60 years. In total, more than 40,000 measurements have been performed. Two critical periods in accumulation of bone-seeking contaminants have been found: (1) period of juvenile growth spurt at the age of 12–15 years, when both incorporation and long-term retention of contaminants is higher by a factor of 4–5 compared to mature adults or younger children; and (2) period of pregnancy and lactation, when contaminant accumulation is higher by a factor of 1.5–2 compared to women of reproductive age. Many residents (about 800 persons) were measured for their ^{90}Sr content for three or more times over a long-term time interval (5–24 years); thus, these extensive data allow evaluation of cortical bone-resorption rate, because resorption by osteoclasts is almost the only way for elimination of the contaminant from the cortical bone. Individual estimates of cortical bone-resorption rate have been obtained for each person in the study that included men and women aged 30–85 years. It was found that rate of resorption is independent of age in adult men aged less than 55 years and in adult women before menopause. In this age period, the rate of cortical bone resorption is independent of gender and close to 3%/year. In women after menopause, the rate of cortical bone resorption depends on time since the onset of menopause. The rate increases by a factor of 2–2.5 compared to pre-menopause and remains at this level for 10–12 years; after that, there is a statistically significant decline in the rate of cortical bone resorption. In men after 55 years of age there is a gradual age-dependent increase. Statistically significant negative relationships were also found between cortical bone-resorption rate and ^{90}Sr-body burden or dose absorbed in skeletal tissues sensitive to radiation exposure [the red bone marrow and the cells lining endosteal surfaces]. However the effect of ^{90}Sr-radiation exposure is insignificant for most of the persons included in the study of cortical bone-resorption rates. As a whole, the effect of ^{90}Sr-radiation exposure on the cortical bone-resorption process was less than the effect of age or menopause in ageing persons.

Characteristics of the Structure of Cortical Bone and Accumulation of Contaminants

Two morphological types of bone are distinguished: cortical (compact) and trabecular (spongy) bone. They differ in strength, presence of the cavities and soft tissues. Cortical bone forms the outer wall of the skeletal bones and the diaphysis of long bones. Trabecular bone forms a system of trabecular beams between the plates of flat bones (scapula, skull), it also forms a network structure within the epiphysis of long bones, and it is a major structural formation in the vertebral bodies, etc. In this regard, it is difficult to uniquely classify a particular bone as cortical or trabecular, because any bone, except the temporal, contains both types of bone tissue. Usually the terms "predominantly cortical" (for example, femoral or tibia bone) and "predominantly trabecular" bone (for example, sternum or vertebrae) are used.

In general, the weight ratio of cortical and trabecular bone in the human skeleton depends on age (in infants all bones can be considered as trabecular) and for an adult, it is about 4:1, i.e., 60-80% of the skeletal mass is the cortical, most durable bone (ICRP 70, 1995). With aging there is a loss of bone minerals, which is more pronounced in the trabecular bone; thus, in older people the percentage contribution of cortical tissue increases with age.

An important quantitative indicator of metabolic processes in the bone is the specific surface area of the bone, on which the cells participating in the process of remodeling (lining cells, osteoblasts, osteoclasts) are located. The ratio of bone surface to volume is about 30 cm^2/cm^3 in cortical; and about 180 cm^2/cm^3 in trabecular bone, i.e., the specific area of trabecular bone is about six times greater than that of cortical. This leads to a significantly lower rate of cortical bone remodeling, and hence, low plasticity (ICRP 70, 1995). The specific area also reflects the contact area with the bloodstream through which the metabolism of the bone tissue occurs.

Exchange processes in the bone are related to ion exchange on the surface of hydroxyapatite crystals. The size and properties of the crystals vary dependent on bone type, age and other factors. It was shown that the surface of crystals/minerals constitutes 1-2 m^2 per g of bone (Driessens et al., 1990). The contact surface bone matrix/crystal is significantly large and may amount to 100–200 m^2 per g of bone. Due to the large surface of hydroxyapatite crystals, the bone tissues have unique ability to absorb different elements and substances from circulating blood (Driessens and Verbeek, 1990; Zaichick et al., 2010). It is supposed that Na, Mg, F, citrates and the most part of CO_3^{2-}

included in minerals are absorbed on the surface of crystals. Some radionuclides, including radium (Ra) and plutonium (Pu) isotopes are also absorbed on the crystal surfaces.

Metals occurring in blood circulation may replace Ca in crystal lattice of mineralized bone. Such metals include strontium (Sr), barium (Ba) and magnesium (Mg). These metals strongly fix in the bone and can only be eliminated from the bone as a result of bone and crystal resorption by osteoclasts. Therefore, the bone tissue serves as a depot of mineral elements and contaminants.

Bone-seeking 'calcium-like' elements include Sr, Ba, and Ra. Lesser analogies of Ca are lead (Pb) and uranium (U). Some of these elements are toxic even in very small amounts (for example, 1 mg of lead produces a toxic effect), other elements have a number of radioactive isotopes that can be ingested or inhaled. Intakes of radioactive isotopes by a mother before or during pregnancy and breastfeeding result in their transfer to the fetus and infants via breast milk resulting in radiation exposure of a suckling baby. Similar patterns are characteristic for other bone-seeking contaminants that may produce toxic effects in the fetus and suckling infant. Bone-seeking radionuclides (for example, ^{89}Sr, ^{90}Sr, ^{140}Ba, and ^{226}Ra) incorporate in the skeleton and irradiate radiosensitive cells within the bone tissues (so called bone surfaces, BS) and the red bone marrow (RBM).

Strontium can be considered as the most studied 'Ca-like' element. Strontium is a trace element, but its content in the human body is quite large, about 320–500 mg (ICRP 23, 1975; Staub et al., 2003 a,b). About 99% of the total strontium-body burden is localized in the skeleton (Staub et al., 2003; Cabrera et al., 1999). The concentration of strontium in food varies in a wide range: In meat, vegetables and fruit the concentration of strontium may vary from 0.3 to 5.1 mg/kg; in grain and seafood – up to 25 mg/kg. Content of strontium in individual diet varies depending on the geographic area and dietary preferences (Cabrera et al., 1999; Tandon et al., 1998). Experiments with rats and guinea pigs showed that at very low strontium content in their diet a defective mineralization of bones and teeth occurred suggesting that strontium is a necessary element for normal development (DNC 2008). On the other hand, excess of strontium and barium and simultaneous deficiency of calcium in the diet result in defects of bone mineralization, deformation of joints and even strontium rickets in children and adolescents. These symptoms (Kashin-Bek disease or endemic osteoarthritis, strontium rickets) are observed in some geochemical provinces in China, Central Siberia and foothills of the Pamirs (Tajikistan).

It is important to note that there is discrimination between strontium and calcium resulting in less intensive transfer of strontium, relative to calcium, through biological barriers: gastrointestinal tract, placenta or breast alveoli (Wasserman 1998). The discrimination against other bone-seeking elements relative to calcium is more enhanced than that of strontium due to larger atomic radii of these elements (Ba, Ra, etc.).

The more important radioactive strontium isotopes are anthropogenic ^{89}Sr and ^{90}Sr[1]. They occur in wastes at reprocessing of spent nuclear fuel and have been released by explosions of nuclear weapons. As a result of testing of nuclear weapons in the atmosphere, which took place from 1945 until 1980, with maximum in 1961–1962, radioactive materials directly released in the environment resulted in global radioactive contamination of the biosphere with radioactive isotopes, including ^{89}Sr and ^{90}Sr. Most of the atmospheric tests were conducted in the northern hemisphere; for this reason, deposition of radionuclides was greater there than in the southern hemisphere. The total deposition of ^{90}Sr in the northern hemisphere from nuclear weapon testing amounted to about $4.7 \cdot 10^{17}$ Bq (UNSCEAR 2000). Thus, ^{90}Sr from global fallouts can be considered as an ecological factor influencing the worldwide population over a long time.

Release of radioactive materials into the environment also occurred as a result of different radiation incidents and accidents at enterprises and plants that use nuclear technologies. The largest radiation accident occurred at the Chernobyl Nuclear Reactor in 1986 in Ukraine that resulted in a severe release of radioactivity into the environment. Large areas in Ukraine, Belarus, and Russia were contaminated by ^{90}Sr, ^{137}Cs and other radionuclides. The total amount of ^{90}Sr released to the environment amounted to about $0.1 \cdot 10^{17}$ Bq (UNSCEAR 2000).

Less well known is radioactive contamination of the small Techa River in the Southern Urals in Russia (Figure 1). The contamination occurred in 1949–1956 as a result of releases of radioactive materials from the first Russian site for production of weapons-grade plutonium, the Mayak Production Association (Vorobiova et al., 1999). About 114 PBq of activity were released in the river; 98% of the total activity was discharged in 1950–1951 (Degteva et al. 2011). Ingestion of large amounts of ^{90}Sr by residents of the Techa Riverside villages occurred with the river water and cow's milk. Since mid-1951 extensive radiometric and radiochemical investigations of the levels of

[1] Strontium-90 is a long-lived radioactive isotope with a half-life of 28.8 years. Strontium-89 is a short-lived radioactive isotope with a half-life of 50.5 days.

radioactive contamination of humans and the environment have been performed in this area. The results of numerous measurements of ^{90}Sr in humans and the environment have been collected and systematized in the databases established at the Urals Research Center for Radiation Medicine (URCRM, Chelyabinsk, Russia).

These data were used for different aims, including reconstruction of the doses for the affected population (Degteva et al. 2000a,b; 2006) necessary for estimation of health effects in the Techa River residents (Krestinina et al. 2005, 2007, 2010); study and modeling of strontium metabolism in humans (Tolstykh et al. 1997, Degteva et al. 1994; Shagina et al. 2003a,b), including strontium transfer to the foetus and breast milk, (Tolstykh et al. 1998, 2008; Shagina et al. 2007); and estimation of cortical bone resorption rates (Shagina et al. 2012, Tolstykh et al. 2011a).

Due to the high level of contamination of river water and floodplain soils, the residents of 27 of 41 settlements on the Techa River were relocated to non-contaminated territories by 1961. The intake schedule for evacuated and non-evacuated persons can be described as chronic intake with peak intakes in 1950–1951 (more than three orders of magnitude higher than in the subsequent years).

Figure 1. A photograph of the Techa River and a typical village located on its shores (photo was taken in the 2000s). The view of houses and households didn't change significantly since the period of the contamination. At present, the river water is relatively clean, while the floodplain soils and the river bottom sediments are still contaminated.

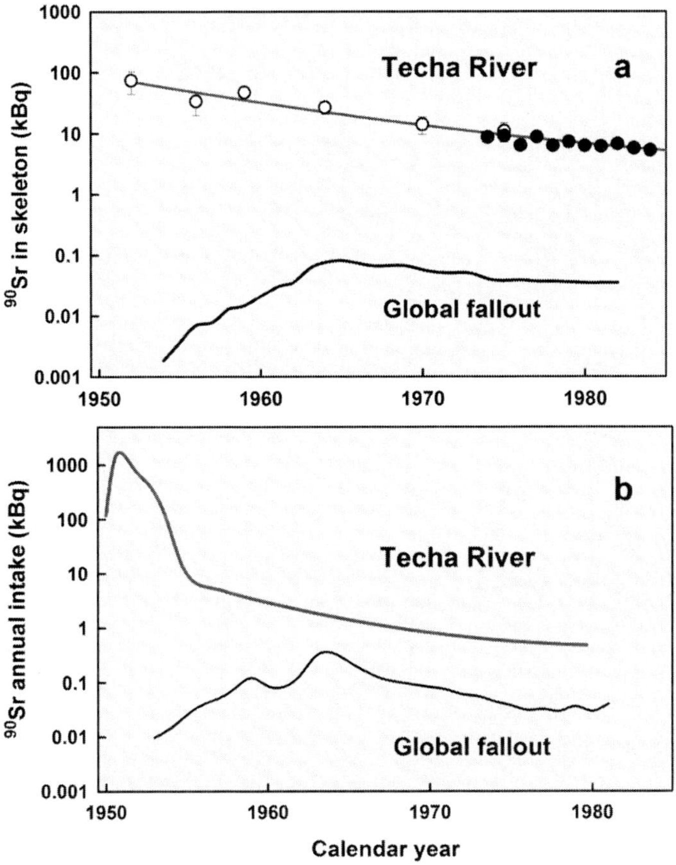

Figure 2. (a) The 90Sr-body contents in upper- and mid- Techa River residents in comparison with contents due to global fallout (Boecker et al., 1991). (b) The 90Sr-intake function for residents of upper- and mid- Techa River for the example of reference settlement in comparison with global fallout data (Papworth and Vennart, 1984).

Figure 2a illustrates a time dependence in ^{90}Sr-body contents in adult residents who lived in the upper Techa River region and levels of ^{90}Sr in the skeleton in adult UK residents due to global fallout. The ^{90}Sr-dietary intake function in a reference Techa River village (Tolstykh et al., 2011b) is presented in Figure 2b derived from ^{90}Sr measurements in human teeth and body, data on radioactive releases (Degteva et al., 2011), and results of ^{90}Sr measurements in environmental samples. As can be seen from Figure 2, both the ^{90}Sr body-burden and intake due to global fallout is insignificant in comparison with that for Techa Riverside residents.

Techa River Strontium Data on Cortical Bone

Table 1 presents available data sets on ^{90}Sr measurements in the Urals residents including people from the Techa Riverside settlements; these data were partly used in Figure 2. Radiochemical ^{90}Sr measurements in different bones provide information on ^{90}Sr distribution in among human bones. The most extensive measurements were obtained with a Whole Body Counter (WBC) that allows *in vivo* measurements of ^{90}Sr-body burden in humans. *In vivo* measurements for the Techa River residents commenced 25 years after the maximum ^{90}Sr intake and continue until the present time. The contents of ^{90}Sr are determined by measuring the bremsstrahlung of the ^{90}Y-beta rays with a phoswich detector. Two modifications of the WBC system (SICH-9.1 in 1974–1997; and SICH-9.1M in 2006–2012) have been used at the URCRM over a long-term period (Kozheurov et al. 2002; Bougrov et al. 2008). All data on ^{90}Sr measurements are kept in a specialized URCRM database.

WBC measurements performed at later periods after the major ^{90}Sr intake (25–60 years after) reflect ^{90}Sr content in cortical bone. The reason why the WBC data can be used for study of the cortical bone is discussed below.

It is well known that at short time periods after ingestion, ^{90}Sr (as well as other bone-seeking contaminants) mainly accumulates in trabecular bone due to a high rate of trabecular bone remodeling and its high ratio of bone surface to volume (Marshall 1972; Papworth and *Vennart* 1984). However, the same reasons result in more rapid ^{90}Sr elimination from the trabecular bone in comparison with the cortical bone. Therefore, at late periods after intake, ^{90}Sr is detected predominantly in cortical bone. This phenomenon was shown in animal experiments (Lloyd et al. 1976; Lubashevsky 1980) and in the study of bone samples contained at the URCRM Autopsy Registry (Tolstykh et al. 2000; Shagina et al. 2003a). This Registry contains data for only a few bone samples from persons who died during the period of radioactive releases, and for only one person (identification code IC 3664) both trabecular and cortical bones were measured. Table 2 exemplifies the ^{90}Sr distribution in different types of bone (i) just after the main ^{90}Sr intake, i.e., in 1952 (IC 3664); (ii) 8–10 years after (IC 99965; 291375; 118846); and (iii) more than 20 years after the main intake of ^{90}Sr (IC 322369, 154009, 143957). All persons lived in settlements on the Techa River from the beginning of radioactive releases up until their deaths. As seen in Table 2, one year after the main ^{90}Sr intake the concentration ratio of predominantly cortical (femur) to trabecular (sternum) bones was about 0.1. The ratio increases with time and becomes equal to 1.7–4.0 after 8–10 years. In late times the ^{90}Sr concentration in samples composed of predominantly cortical bone are higher than ^{90}Sr in trabecular bone by a factor of 12–24. Therefore, the cortical bone can be assumed as a “long-term storage” of bone-seeking contaminants. Figure 3 illustrates ^{90}Sr distribution in different type of human bone at more than 20 years after ^{90}Sr intake.

Table 1. Data sets on ^{90}Sr measurements in Urals residents

Method of measurements	Number of persons	Number of measurements	Period of measurements
In vivo measurements with the WBC systems	25,000	43,000	1974-2012
Radiochemical and radiometry measurements of bone samples obtained post mortem	6,000	10,250	1951-1989

Table 2. Examples of ^{90}Sr concentration in different samples of human bone (according to Tolstykh et al. 2011a with modifications)

Identification code	Time after ^{90}Sr intake, years	Number of bone samples	^{90}Sr concentration ratio of cortical to trabecular bones[#]
3664	1	2	0.11
99965	8	4	1.68
291375	8	4	3.88
118846	10	5	4.01
322369	20	4	12.0
154009	25	3	16.7
143957	30	3	23.9

[#] Femur, fibula, and cranium were considered as predominantly cortical bones; vertebra and sternum were considered as predominantly trabecular bones.

Therefore, almost the entire amount of ^{90}Sr had been deposited by 25 years following intake in the cortical part of the skeleton (built into crystals of hydroxyapatite). Because cortical bone accounts for up to 60–80% of skeletal weight in adults (ICRP 1995), it can be assumed that essentially all strontium measured with the WBC at this time after intake is contained in the cortical part of the skeleton (shaft of long bones).

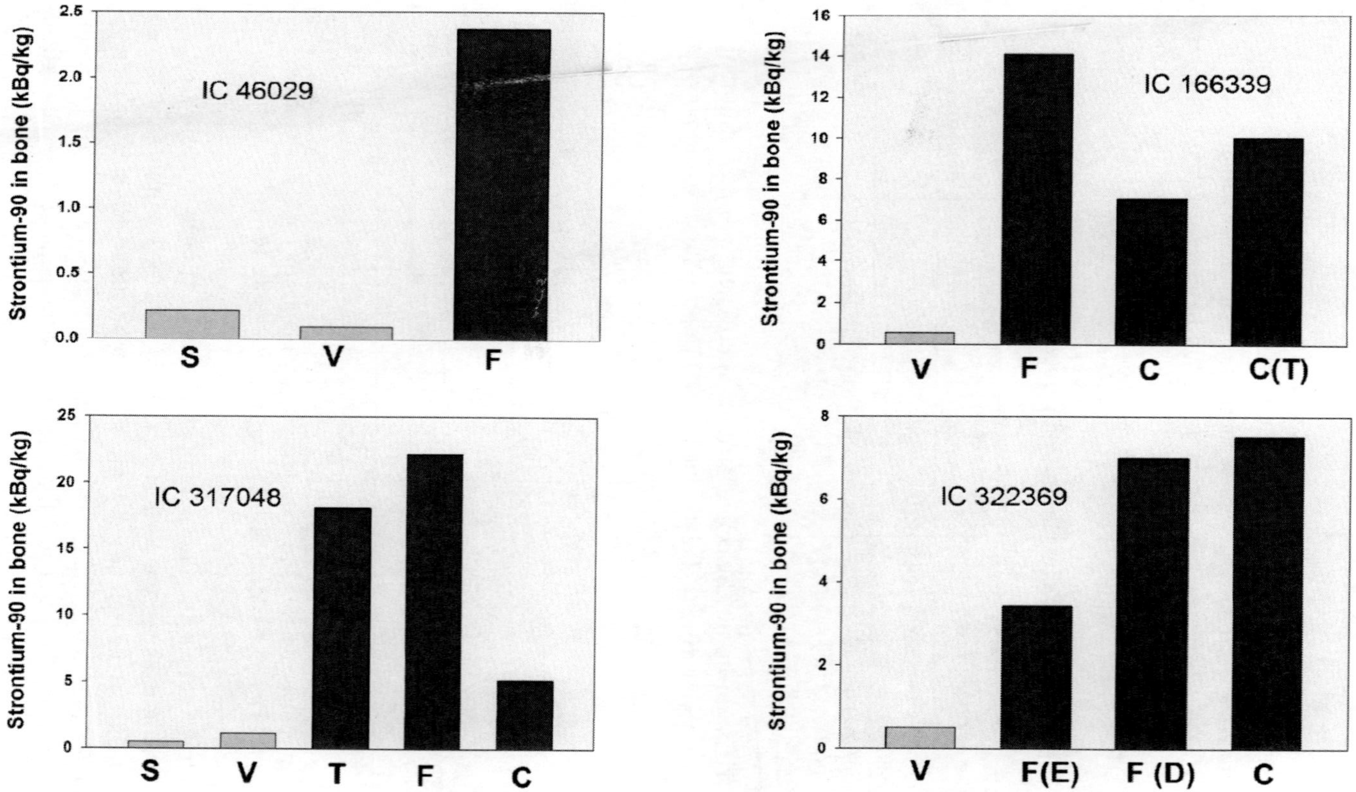

Figure 3. (Continued).

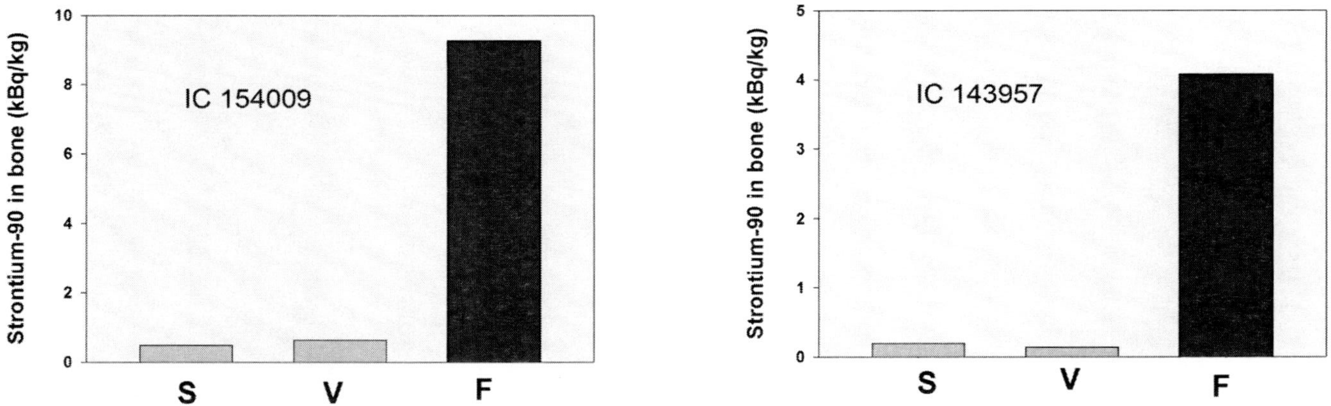

Figure 3. Examples of ^{90}Sr concentration in different human bones. Dark bars indicate predominantly cortical bones. IC – individual identification code in URCRM data bases. The following bones or bone sites are shown: Sternum (S), Vertebra (V), Fibula (F), Epiphyses (E), Diaphysis (D), Tibia (T), Cranium (C), and Temporal bone C(T).

Cortical Bone of Adolescents is the Best "Long-Term Storage" for Toxic Contaminants

Figure 4 represents unique data on long-term ^{90}Sr retention in the cortical bones of males and females of different age at the time of radionuclide intake. As can be seen, ^{90}Sr is well detected both 30 and 45 years after major intake.

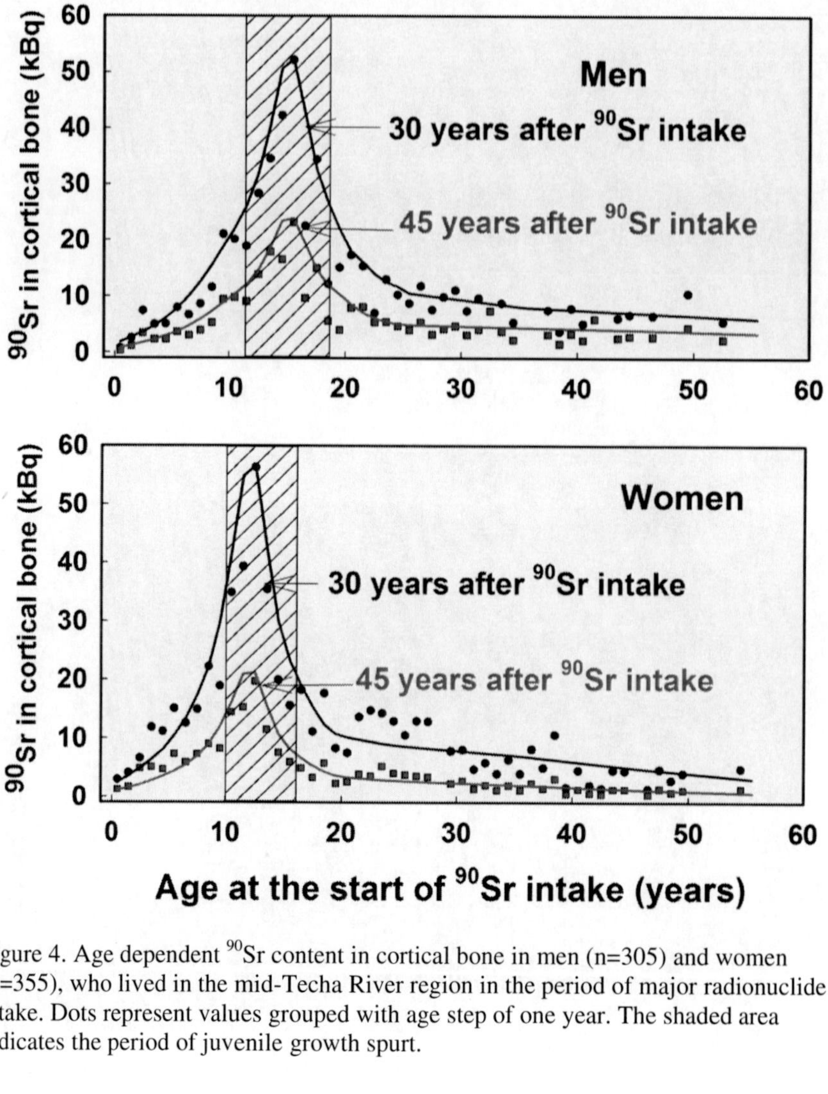

Figure 4. Age dependent ^{90}Sr content in cortical bone in men (n=305) and women (n=355), who lived in the mid-Techa River region in the period of major radionuclide intake. Dots represent values grouped with age step of one year. The shaded area indicates the period of juvenile growth spurt.

Data on measurements performed in the 2000s are not so numerous. This is explained by natural reasons, because a significant numbers of people living in the Techa Riverside settlements in 1950s have already died. However, even 60 years after the major ^{90}Sr intake significant amounts of ^{90}Sr are detected in cortical bones of people who were in their adolescence at the time of the intake (Figure 5).

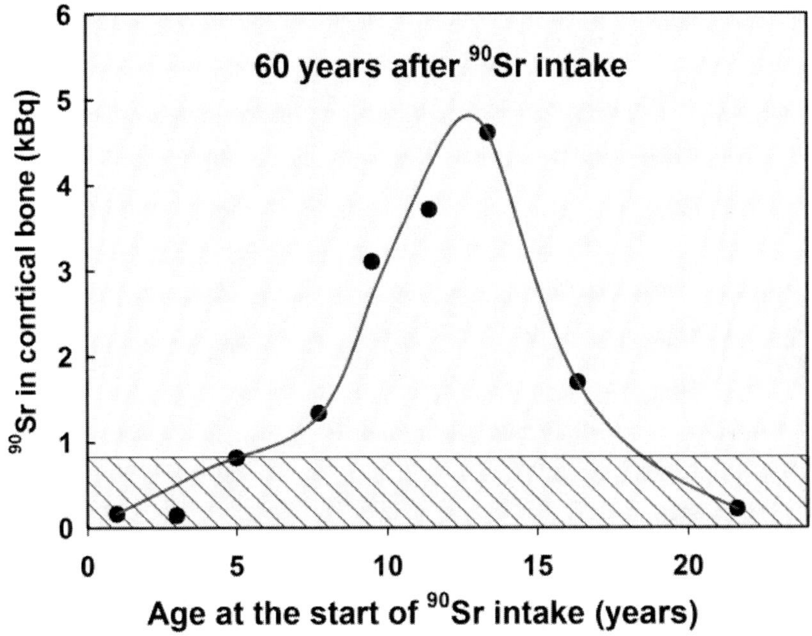

Figure 5. Age dependent ^{90}Sr content (median values) in cortical part of the skeleton combined for men and women, who lived in the upper Techa River villages during the major radionuclide intake; n=386. The shaded area indicates levels below WBC detection limit.

The observed age- and gender-dependences of ^{90}Sr in the cortical bones have a clear physiological interpretation. The following facts are of the main interest:

1. Ingested Sr (and Ca) accumulates in the skeleton in the sites of growth and calcification.
2. Maximum rate of skeletal and height growth occurs 1–2 years before puberty (Donoval et al. 1965). There is a strong correlation between individual skeletal age and menarche age in girls (Simmons et al.

1943; Abrams et al. 1996). Analysis of available data on gynecological examinations showed that the average age of menarche in Techa settlements in 1950^s was 15 years.

3. The pubertal growth in boys occurs 2–2.5 years later compared to girls. As with girls, maximum skeletal growth, particularly of the cortical bone precedes the end of puberty. In this period, maximum retention of ^{90}Sr occurred as shown in Figure 4. An age shift in maximum ^{90}Sr accumulation in the cortical bone in boys compared to girls is clearly seen from Figure 4.

The fact that maximum accumulation of contaminants occurs in adolescents is confirmed by data from studies of global fallout. This is demonstrated in Figure 6 showing ^{90}Sr concentration in the rib bone of males and females of Chelyabinsk Oblast (Southern Urals, Russia) 10 years after peak dietary ^{90}Sr intake from global fallout. For comparison, data obtained in Western Germany 13–16 years after the peak dietary ^{90}Sr intake from global fallout are presented in Figure 6b (Dehos at al. 1985; original data were recalculated relative to 1964). Statistically significant differences between ^{90}Sr concentration in males and females were not found.

As can be seen from Figure 6, the highest levels of ^{90}Sr retention in the skeleton were observed in persons who were in their adolescence during peak dietary ^{90}Sr intakes from global fallout in 1964. It should be noted that the levels of ^{90}Sr concentration in bones from the Urals and German populations differ significantly.

This is explained both by the time of bone sampling (or time periods after the peak intakes) and the type of bone used in analysis. In Western Germany analysis of ^{90}Sr concentration was performed in samples of femur diaphysis, which represents the most mineralized cortical bone with a very low rate of bone turnover.

Therefore, the cortical bone can be used as an indicator of environmental contamination by bone-seeking contaminants. Cortical bone reflects the level of water and foodstuffs contamination during the period of enhanced skeletal growth, or at the age of 12–15 years old. Contaminants accumulated in the skeleton during the adolescence can be detected in the cortical bone decades afterwards.

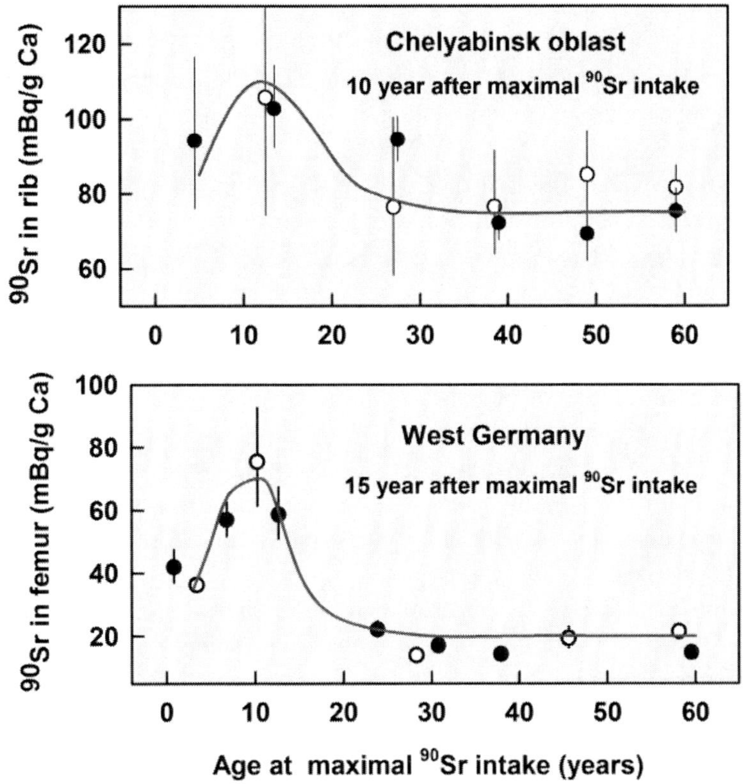

Figure 6. Concentration of ^{90}Sr in bone samples for males (black circles) and females (white circles) of different ages during peak ^{90}Sr dietary intake due to global fallout in the southern part of Chelyabinsk Oblast and Western Germany. Data on Chelyabinsk Oblast represent ^{90}Sr concentration in rib samples, taken in 1973–1975, n=361; data on Western Germany represent ^{90}Sr concentration in samples of femur diaphysis taken in 1977–1980, n=118 (Dehos at al. 1985). Geometric means and geometric standard deviations are presented.

Cortical Bone of Pregnant and Breastfeeding Women Has the Second Largest Impact on Accumulation of Toxic Contaminants

Changes in mineral metabolism during pregnancy are necessary to meet requirements of a growing fetus (Kovacs et al. 2006). These changes include an increase in calcium absorption from the gastrointestinal tract (Cross et al. 1995, Fell et al. 1998, Ritchie et al., 1998), increase in urinary calcium excretion (Heaney et al. 1971, Kovacs et al. 2006, Pitkin 1985) and increase in bone turnover although interpretation of data on bone turnover markers is ambiguous (Kovacs et al. 2006). Data from longitudinal studies show that pregnancy does not result in net mineral loss from the maternal skeleton; however, the bone turnover rate during pregnancy increases almost two-fold. Such an increased bone turnover remains during breastfeeding and a decrease in mineral density of trabecular bone has been noted (Krebs et al. 1997; Cross et al. 1995; Richie et al. 1998; Heaney et al. 2003), which returns to normal physiological levels only after the resumption of menses. These indicate sharp changes in accumulation of bone-seeking elements in the maternal skeleton during pregnancy and breastfeeding. However, quantitative evaluation of the outcome, i.e., of the accumulation of bone-seeking elements in the maternal skeleton, particularly in the cortical bone, is practically impossible with the use of standard methods. Data from the Techa River studies provide a unique possibility to quantify this parameter.

Figure 7 presents WBC measurements of ^{90}Sr in the cortical part of the skeleton in women who lived in the Techa River villages during major ^{90}Sr-dietary intake. The results of WBC measurements were averaged in larger age-groups in comparison with data presented in Figure 4. As a result, increased ^{90}Sr content in women aged 21–23 years, which looks as a statistical variation in Figure 4, becomes significant (shown in Figure 7). The reason for such increased levels of ^{90}Sr in these women was not clear for a long time. It was only found in our recent studies on modeling of calcium and strontium metabolism in pregnant and breastfeeding woman (Shagina et al. 2007, Tolstykh et al. 2009) that in this group of females most women were pregnant or breastfeeding. In closer age groups of females only from 25 to 50% of women were parous.

For a detailed analysis, WBC measurements were selected for 100 women who lived in the same settlement on the Techa River and were 19–29 years old at the beginning of ^{90}Sr intake (born in 1921–1931). Three groups were selected from them according to their child-bearing history.

- The first group included women who had successful child-bearing in the period of maximal ^{90}Sr intake (1950–1951); women who gave birth to one or two successive children;
- The second group included women who gave birth *after* the period of maximal ^{90}Sr intake (1952–1961); and

- The third group comprised women who gave birth before the period of ^{90}Sr intake or were childless.

The results of WBC measurements for these women were normalized to a fixed date (1980). As can be seen from Figure 8 women who gave birth in the period of maximal ^{90}Sr intake have higher ^{90}Sr content in cortical bone by a factor of 1.5–2 than women who gave birth after or before this period or did not have children. The differences between group 1 and group 3 are of the main interest, because the women from group 2 received some amount of ^{90}Sr with diet during pregnancy/lactation; and differences by a factor of 2 obtained for group 3 should be assumed as more realistic. The reasons for increased ^{90}Sr-accumulation in cortical bone are the following:

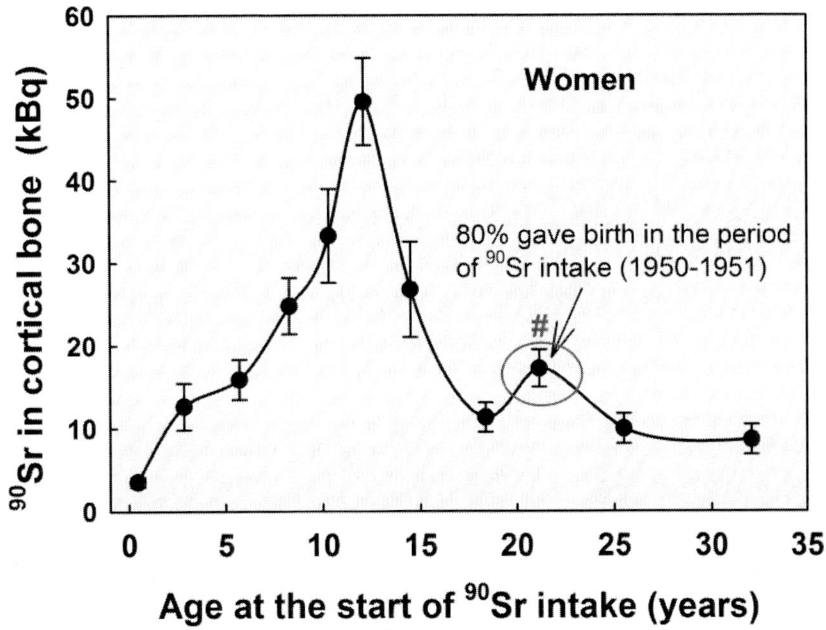

Figure 7. Age-dependent ^{90}Sr content in cortical bone in women who lived in the middle Techa River region during the period of major ^{90}Sr dietary intake.

Increase in ^{90}Sr dietary intake in the period of pregnancy and lactation (Tolstykh et al. 2008). It was shown in different studies that even if energy and protein intake before conception were sufficient to meet the demands of fetal development and milk production, women nevertheless show increased consumption of foodstuffs during pregnancy and lactation.

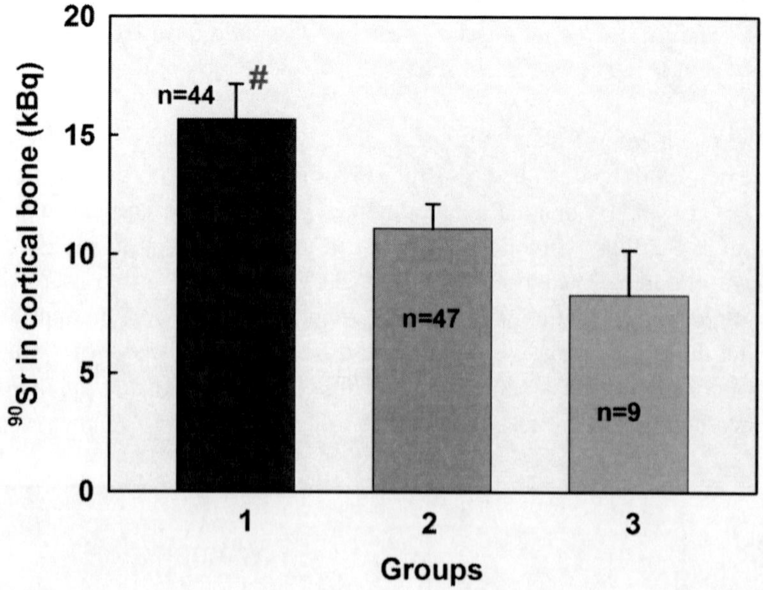

Figure 8. ^{90}Sr contents in cortical part of skeleton measured in women who were pregnant and breastfeeding in different time periods relative to the maximum ^{90}Sr dietary intake. Asterisk indicates a significant statistical difference ($p<0.05$) from other two groups of women as confirmed by Student's t-test and Mann-Whitney U-test.

Increase in bone remodeling, both formation and resorption, is observed in the period of pregnancy (especially in the third trimester) and during lactation, that lead to a detectable decrease of bone mineral density (Cross et al. 1995; Kovacs and Chada 2006; Ritchie et al. 1998).

Rapid Ca/Sr accumulation for restoration of bone after pregnancy and lactation. It was shown that bone-mineral density rapidly returns to pre-pregnant levels after the restoration of hormonal level and renewal of menstrual cycle.

Therefore, women in the period of pregnancy and breastfeeding accumulate significant amounts of contaminants. Interestingly, their children accumulate contaminants from transfer from the maternal body during gestation as well as from breast milk and infant diet during the first years of life. However, a high rate of skeletal turnover during childhood results in rapid elimination of contaminants from the skeleton. For example, for children born after the beginning of ^{90}Sr intake in the Techa Riverside villages, the levels of ^{90}Sr in the skeleton were low and were not detectable with the WBC.

Cortical Bone-Resorption Rate Estimated from the Techa River Data

The resorption of cortical bone can be considered as the following chain of events (Manolagas 1995; 2000; Parfitt 2006; Teitelbaum 2000; 2007): Osteocytes communicate via the canaliculi with lining cells (inactive osteoblasts or a cell type of its own) on the nearest bone surface and give a signal for local bone-area remodeling.

As a result, the retraction or removing of lining cells occurs in an area of local bone. Retraction exposes the underlining osteoid to proteolytic enzymes of osteoblasts, and further to osteoclastic resorption. The local bone-area code for circulation of mononuclear osteoclast precursors is turned on. Osteoclast precursors leave the circulation and migrate to the exposed mineral, where they fuse into multinucleated osteoclasts and begin to resorb bone. In normal tissue the resorption rate can be regulated in two ways, either affecting the number of osteoclasts formed or the resorption power of an individual, mature osteoclast. Both of these mechanisms have been shown to be effective in determining the resorption rate, and they can either be controlled by local or systemic regulators (Lorenzo et al. 2008; Manolagas 1995; 2000; Parfitt 2006; Teitelbaum 2000; 2007).

It is very difficult to separate effects of bone formation and resorption, because the main measured value is the bone-mineral loss/addition which is the net result between bone formation and bone resorption. The use of a radioactive tracer is the best way to separately estimate these processes. Data on ^{90}Sr retention in Techa Riverside residents give us a unique opportunity to quantitatively estimate the resorption rate in the cortical part of the skeleton as a whole.

Strontium-90 can only be excreted from cortical bone due to the process of bone-tissue resorption (*Engström 1958;* Marshall 1972*)*. Therefore, the rate of biological elimination of strontium depends directly on the rate of cortical bone remodeling.

Unique Method of Evaluation of the Cortical Bone-Resorption Rate with the Use of ^{90}Sr as a Tracer

Many Techa Riverside residents were repeatedly measured for their ^{90}Sr-content in the cortical bone over a long-term time period. Some of them were measured more than 10 times during 35 years. The measured persons were included in the study of cortical bone resorption under the following conditions. A person had to have had three or more WBC measurements over a period of five years or longer. If a person was measured only three times, the interval between two successive measurements had to be no less than two years. The results of measurements were considered only if the first three measurements were higher than the minimal detectable activity.

Analysis of repeated WBC measurements has shown that the rate of ^{90}Sr diminution in the human body during periods 25–45 year after major intake can be described by an exponential function (Kozheurov 1994, Shagina et al. 2003a). For each person (total number of studied persons was 777), who had repeated WBC measurements, an exponential function was fit to repeated WBC measurements and the hypothesis of correspondence of an exponential model to experimental data was confirmed statistically with use of the F-test. Figure 9 shows examples of individual ^{90}Sr measurements for a man (Identification Code (IC) 101299) and a woman (IC 95860) and exponential functions fitted to individual WBC data.

It is obvious that the rate of ^{90}Sr diminution and the rate of cortical bone resorption are numerically different, because some fraction of resorbed minerals is returned into the bone surface and into the sites of bone remodeling. Therefore, the estimation of bone-resorption rate from repeated WBC measurements requires evaluation of the rate of strontium recirculation.

Strontium recycling in the human body can be described in the following way (Bronner et al. 1963; Bronner 2000). As a result of osteoclast activity, some portion of hydroxyapatite crystals from the bone matrix goes into solution. Calcium, strontium, and phosphate ions migrate to blood vessels through the interstitial fluid.

There is ongoing ion exchange on the surface of the mineralized bone matrix, and a portion of calcium and strontium ions, transferred with blood flow, precipitates there. This process occurs in both directions, i.e., from the bone surface and to the bone surface.

On the bone surface the processes of bone resorption and bone formation are always taking place. Part of the strontium in the blood returns to the sites of bone formation and is fixed there, and the non-fixed part is excreted from the body with urine (about 70–75%) and feces (25–30%) (Legget 1992; Lenihan et al. 1967; Likhtarev et al. 1975). Formally, the process of ^{90}Sr-elimination rate and cortical bone-resorption rate can be described as follows:

$$\lambda_b = \lambda_{cb} - \lambda_r ,$$

where λ_b is the strontium biological elimination rate, i.e., diminution rate minus the rate of radioactive decay (2.4% per year), λ_{cb} is the cortical bone-resorption rate, and λ_r is the strontium-recycle rate.

It is difficult to estimate the actual value of strontium recirculation, because it is problematic to elaborate an adequate experimental scheme for its determination. In our study the rate of strontium recycling was derived with use of the biokinetic model for strontium recommended in Publication 67 of the International Commission on Radiological Protection (ICRP 1993; Legget 1992). Two kinds of model calculations for the time period of interest (25 years after intake) were performed: 1) calculation based on the full set of model parameters; and 2) calculation with the exception of reverse transfer of strontium from plasma to bone surface. As a result, the annual strontium-elimination rates were calculated. The rate of recycling was determined as a difference between the two values of model estimations.

The value of the strontium-recycling rate was estimated to be 0.18% per year (Tolstykh et al. 2011a). Thus, the values of strontium-elimination rates were corrected in order to calculate the rate of cortical bone resorption. It is obvious that the value of 0.18% per year is a rough estimate. The recycling rate is dependent on the individual features of mineral metabolism and can vary with age. However, the persons in our study were adults by the start of WBC measurements, and the strontium-recycling rate was considered to be a constant.

The derived individual ^{90}Sr-diminution rate was corrected for ^{90}Sr radioactive decay rate (λ_d = 2.4% per year) to receive the biological elimination rate, λ_b. Through use of the equation above, the rate of cortical bone resorption was then calculated for each person. The relative standard errors of individual values of λ_b were in the range of 1–40% with an average value of about 10%.

An averaging procedure allowing estimation of an average cortical bone-resorption rate for a specific age is not evident. Taking into consideration that the estimate of the resorption rate is obtained over a long age-period, it is important to consider to which age an individual estimate should be assigned. The estimate can be assigned to one age, corresponding to a middle age over the whole period of measurements, or it can be assigned to different age categories along the period of measurements. Both approaches were used in our studies for evaluation of average ^{90}Sr-diminution rates. Initially (Kozheurov et al. 1994, Tolstykh et al. 1997) each person was characterized by a *single* value of diminution rate irrespective of age-interval length covered during the period of WBC measurements. In later analysis of repeated WBC measurements (Shagina et al. 2003a) individual ^{90}Sr-diminution rates were considered *several* times in an averaging procedure according to age intervals covered during the period of WBC measurements. This approach was used in subsequent investigations (Tolstykh et al. 2011a, Shagina et al. 2012).

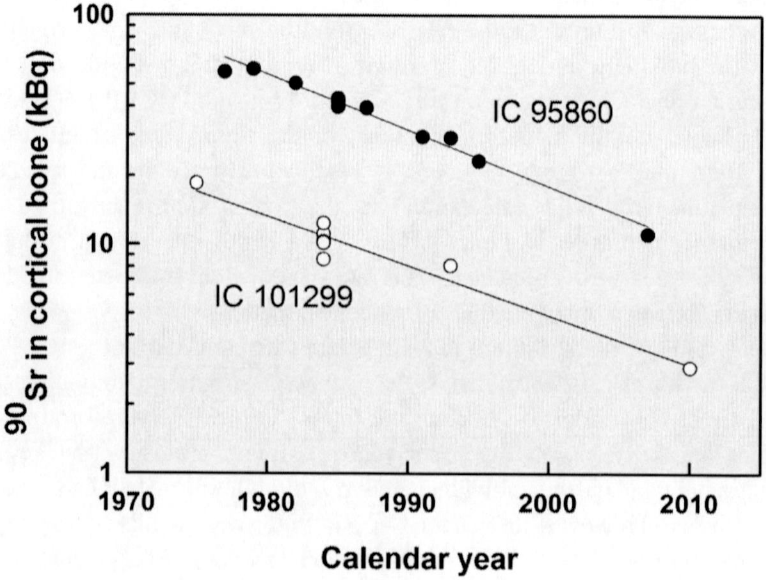

Figure 9. Repeated WBC measurements of ^{90}Sr in cortical bones obtained for two persons from the Techa River settlements. The person identified by IC 95860 has been measured with the WBC from 1976 until 2008; the person identified by IC 101299 had WBC measurements in 1974–2010. Circles represent individual measurements; lines represent exponential functions fitted to WBC data.

The scheme presented in Figure 10 illustrates a method of averaging the estimates of cortical bone-resorption rate for a few persons in order to calculate grouped-average resorption rates for a certain age category. As can be seen from Figure 10, an individual's value of cortical bone-resorption rate can make a contribution to the respective average values for several age categories. These persons were measured with WBC in age interval from 30 to 65 years old. For example, the person identified as IC11830 "makes a contribution" to five different age categories. So, that person's value of resorption rate (2.6% per year) is included in the averaging procedure conducted for each of considered age categories: 30–35, 35–40, 40–45, 45–50, 50–55, 55–60 and 60–65 y.

Cortical Bone Resorption Rate in the Period of Relative Stable Bone Mass from 25 to 50-55 Years

The age criterion for men was that measurements were performed during the age-period from 25 to 50 years. For women the age period from 25 years up to age of onset of individual menopause was required. Data on individual menopause were taken from outpatient cards and case histories containing the results of gynecological examinations.

According to the above conditions, a total of 2,022 WBC measurements for 73 women (713 measurements) and 134 men (1,309 measurements) were included. Most measurements were performed for persons in the third and fourth decades of life.

The youngest person measured was 28 years old. The studied persons were 0.5–20 years old at the beginning of ^{90}Sr intake. Table 3 presents the characteristics of the persons under study in terms of cortical bone-resorption rate. It is important that men and women do not differ substantially in bone-resorption rate. Men and women were found to have very similar distributions of resorption rates; both distributions are well approximated by lognormal distributions. The hypothesis of absence of differences is confirmed by the Wilcoxon-Mann-Whitney U Test ($P = 0.48$). Therefore, the groups of men and women were combined for further analysis.

Persons from the studied sample were repeatedly examined at the URCRM Clinics, due to the high levels of ^{90}Sr accumulations and possible resulting changes in their health status.

It is known that some diseases (or some medicines) affect mineral metabolism and bone remodeling. The relationship between the occurrence of such diseases and bone-resorption rates for persons in the sample was analyzed. The following diseases were considered that occurred before and/or during the period of WBC measurements.

- Diseases that required corticosteroid therapy: Agranulocytosis, hemolytic anemia, bronchial asthma, hemoblastosis, and rheumatoid arthritis (14 cases in total).
- Diseases that influenced mineral metabolism and/or required hormone therapy: Thyroid diseases, urolithiasis, and cholelithiasis (38 cases in total).

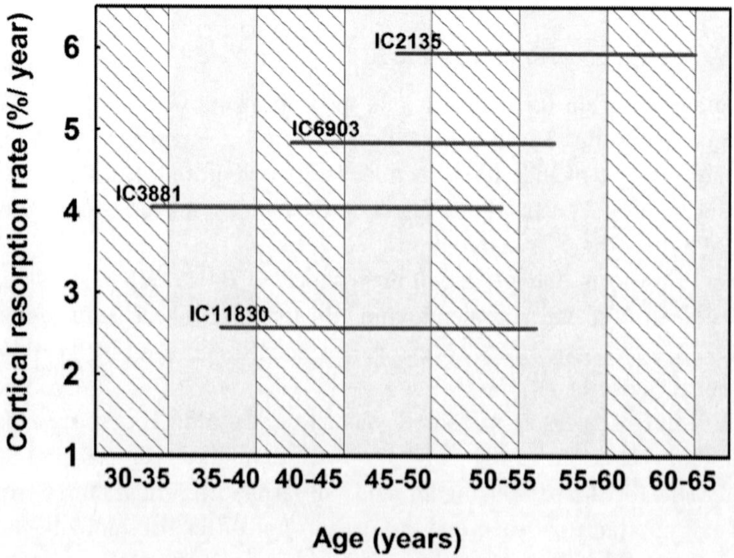

Figure 10. Examples on averaging individual estimates of cortical resorption rates in the considered age categories.

Table 3. Characteristics of persons under study in terms of cortical bone-resorption rate

Statistical characteristics	Cortical bone resorption rate, %/year^{-1}		
	Men	Women	Both
Median	2.8	2.7	2.8
90% range	1.4–5.1	1.2–5.7	1.4–5.2
Min–Max	0.8–9.4	0.6–8.1	0.6–9.4
Geometric mean ±STD	2.8±1.5	2.7±1.6	2.7±1.5

For 10 persons, both diseases from those listed above were diagnosed during the period of interest. The comparison of values of bone-resorption rates in groups combined according to the presence or absence of specific diseases has not shown statistically significant differences between groups.

As discussed in this Chapter, bone remodeling, both formation and resorption, is significantly increased in the period of pregnancy (especially in the third trimester) and during lactation, and then returns to pre-pregnancy values (Cross et al. 1995; Kovacs and Chada 2006; Ritchie et al. 1998). In the sample there were six women who had labor during the period of WBC observation, i.e., they were measured before and after pregnancy and lactation. Table 4 presents information on mothers and their children. As can be seen, there is a tendency for an increased rate of cortical bone resorption in these women. However, the average value is not significantly different from rates of non-pregnant and non-lactating women. So, the issue of pregnancy and lactation for this small number of women was not considered in the following analysis.

Another important question to study in this population is whether the cortical bone-resorption rate changes due to ^{90}Sr-radiation exposure? To answer this question individual doses accumulated in the RBM and BS were calculated according to approaches described by Degteva et al. (2000a,b; 2006; 2007). The RBM dose characterizes the exposure of osteoclast progenitors, which are hematopoietic cells (Parfitt 2006), and some fraction of osteoblast progenitors. [In general, osteoblast progenitors were found in the bone marrow, endosteum, periosteum and soft tissues (Colnot 2009; Schindeler et al. 2009; Watchman et al. 2007)]. The BS dose characterizes the radiation exposure to osteoblasts, some fraction of their progenitors, and lining cells. The average absorbed dose in the RBM was about 1 Gy in men and women.

Table 4. Characteristics of mothers and their children born during the period of WBC measurements

Mother	Mother's year of birth	Cortical resorption rate, %/year	Period of WBC measurements	Date of children's birth
IC 63811	1940	3.2	1977–1988	1979
				1980
IC 91120	1942	3.2	1976–1988	1981
IC 127685	1941	2.9	1974–1987	1975
IC 143618	1939	2.5	1975–1987	1976
IC 209120	1946	2.7	1976–1990	1983
IC 636668	1940	3.3	1976–1987	1978
Geometric mean		3.0 (2.7–3.3)		

Statistically significant negative relationships of cortical bone-resorption rate were discovered related to dose absorbed in the RBM or the BS (Tolstykh et al. 2011a). The response appears to have a threshold of about 1.5 Gy RBM dose. The radiation-induced decrease in bone-resorption rate (0.25% year^{-1} per

1 Gy of RBM dose) may not be significant in terms of health. However, a decrease in bone-remodeling rate can be among several reasons that cause an increased level of degenerative-dystrophic bone pathology in exposed persons (Akleyev et al. 1995).

Figure 11 illustrates values of cortical bone-resorption rates obtained in the present study as a function of average-individual age during the period of measurements.

As can be seen, the rate of cortical bone resorption does not change significantly during the third and fourth decades of life under conditions that women do not reach menopause. Because the values of cortical bone-resorption rates did not depend on ^{90}Sr dose in the range <1.5 Gy, the obtained values can be used for modeling purposes as reference values for adults of reproductive ages.

Cortical Bone Resorption Rate in Men Aged 50–80 Years and for Women 0–30 Years after Menopause

Measurements were used for men aged 50+ years and for women from the beginning of individual menopause. A total of 1,728 WBC measurements of 126 women (1,064 measurements) and 71 men (664 measurements) were included. There were only nine women and five men in the studied group who had the minimum number (three) of WBC measurements; a large number of measurements (above 10) was available for more than one-third of the studied group (42 women and 25 men). The maximum age for men was 83 years and for women 82 years. All studied persons lived in the Techa Riverside villages in the 1950s. Age at major ^{90}Sr intake was 8–38 years old for women and 11–55 years old for men.

To investigate the effect of the diseases described above on the cortical resorption rate, comparison of the distributions for the whole group of persons and for the group of persons with absence of specific diseases was made. It was shown that the presence of specific diseases did not alter significantly the distributions of cortical resorption rates in men or women (absence of statistically significant differences was confirmed by the Mann-Whitney test).

Analysis of the rates of cortical resorption in the group of women showed that the rate was independent on age (in the middle of the period of measurements). There was a tendency toward increased rate of cortical resorption with advancing age of individual menopause; however, this was not statistically significant (Spearman $R = 0.15$, $p = 0.1$).

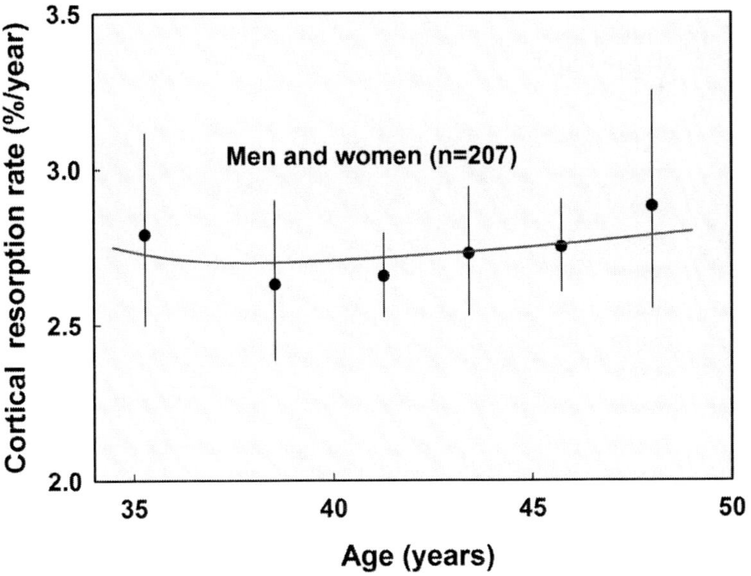

Figure 11. Geometric means of cortical bone-resorption rates as a function of age; combined data for men and women.

Figure 12 shows geometric means of the cortical bone resorption rate in women dependent on the time since menopause and in men dependent on age. It is known that changes in bone resorption in women are connected with a decrease in estrogen levels after menopause (Riggs and Melton ed., 1996; Khosla et al., 1998); thus, age relative to menopause is a more appropriate parameter for women.

It is seen from Figure 12 that the cortical resorption rate in women doubles after menopause and then remains at this level for 10–12 years, after which there is a decline in the resorption rate. The grouped average rates 0–5 and 5–10 years after menopause are statistically significantly larger than the grouped average rates in the older age groups. Therefore, we have formed groups corresponding to two time periods after menopause, 0–10 years and 10–33 years. Geometric means of the cortical bone-resorption rate in these groups are equal to 6.1%/year and 5.4%/year, correspondingly. Differences between the groups are statistically significant ($p<0.001$). It should be noted that in all age categories the cortical bone-resorption rate was significantly higher than the pre-menopausal value, equal to 2.7%/year.

For men (Figure 12), age-dependences in the rates of cortical bone resorption are less pronounced than in women.

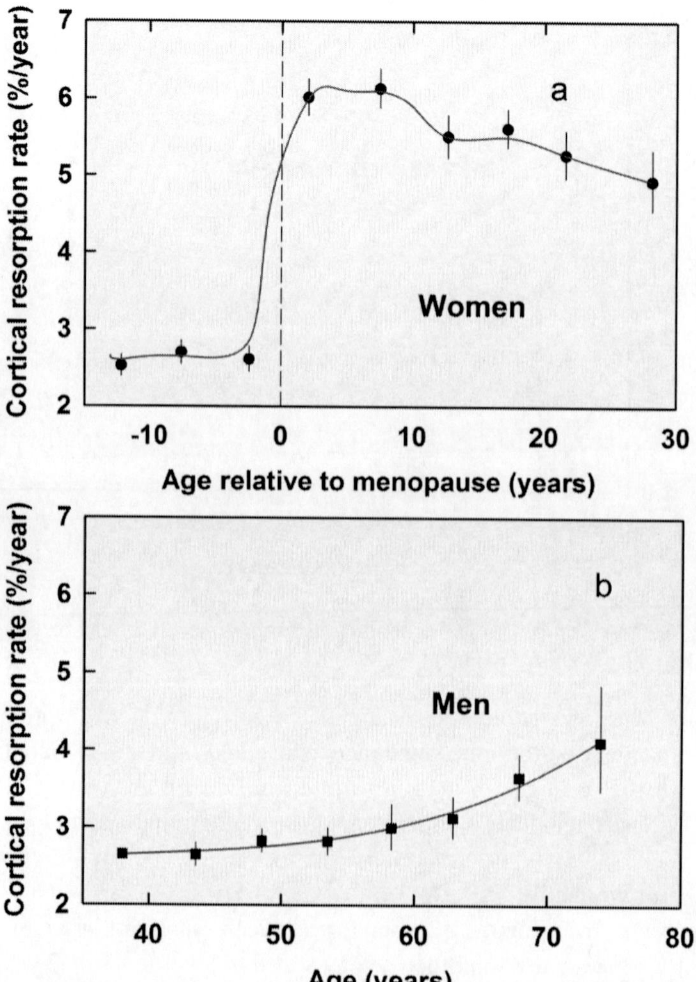

Figure 12. Cortical bone-resorption rate in women dependent on (a) the time since menopause and (b) in men dependent on age. Circles represent geometric means; vertical lines represent geometric standard deviation of individual values.

A statistically significant increase in resorption rate is seen in age groups older than 55 years. There are no statistically significant differences in the cortical resorption rates between the groups in the age interval from 55 to 70 years; however, in the oldest age group (70–75 years) the rate of cortical resorption is statistically different from the younger age groups (55–60 years). These results provide evidence for smaller age-dependent changes in the rate of bone remodeling in men compared to women. It should be noted that there was no age dependency in the rate of cortical resorption in men aged 30–55 years.

Figure 13 shows distributions for the rate of cortical bone resorption in women and men aged 55–80 years. At the age of 55 years, 98% of women from this group experienced menopause, and there were no sharp changes in the cortical bone-resorption after this age. It is seen from Figure 13 that the cortical bone-resorption rate is twice higher in women compared to men over this age interval; this is confirmed by the Mann-Whitney test ($p<0.001$). This is different from what we observed in groups of men and women of reproductive ages (Figure 11).

Therefore, our study allowed evaluation of the rate of endocortical bone resorption in men and women from longitudinal measurements obtained over 24 years of follow-up. Using ^{90}Sr as a radioactive tracer allowed evaluation of the age-dependences in the rate of resorption via an indirect method. Another method that has been used for assessment of cortical bone-resorption rate, on the basis of *in vivo* measurements, is assessment of the cortical bone area and medullary bone area of long bone with peripheral quantitative computer tomography (pQCT).

Estimates obtained in the study presented here are in agreement with qualitative estimations (Recker et al. 2004) and results derived from longitudinal pQCT measurements (Lauretani et al. 2008). These pQCT results showed that age-related medullary expansions in tibia, indicating the expression of endocortical resorption, increased in women after 50–55 years old and in men after 60–65 years old. The widening of the medullary area was considerably steeper in women compared to men over the entire age period (20–100 years). However, in contrast to data of Lauretani et al. (2008) our data did not show a statistical difference in the rate of cortical bone resorption in men and women of reproductive age.

Changes in volumetric bone-mineral density (vBMD) and body-mineral content (BMC) indicate changes in the ratio of bone resorption-to-formation. It is essential to note that cortical bone accounts for 80% of skeletal weight (ICRP, 1995); its contribution increases with advancing age, because the age-dependent loss in trabecular bone is more pronounced than in cortical bone (Riggs et al., 2008). Therefore, BMC loss is mainly determined by the loss of cortical bone mineral.

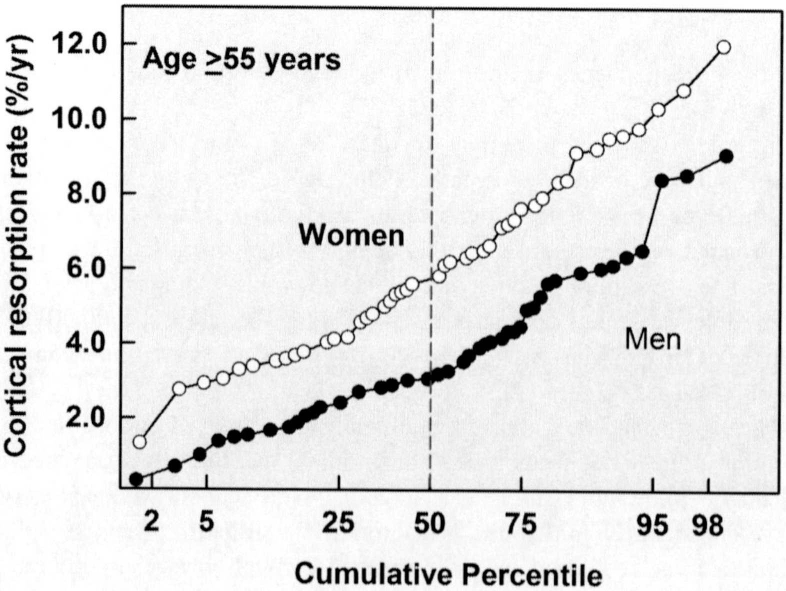

Figure 13. Distributions of the rate of cortical bone resorption in women and men aged 55–80 years.

The age-related changes in vBMD of the cortical bone have been investigated in different studies. The most detailed investigations on the Russian population for the period of interest were conducted by Sveshnikov and Kuznetsov (1989) with Dual Energy Photon Absorptiometry (DPA). Their studies were carried out for the population of Kurgan Region; this population is similar to Techa Riverside residents in terms of life-style. It was shown that vBMD in tibia and femur shafts decreased after mid-life at the rate of 0.4–0.5%/year in men and 0.6–0.8%/year in women. Less pronounced changes in vBMD of distal radius and distal tibia were described for the population of Rochester (Riggs et al., 2008): After 50-years of age the rates were higher in women than in men by a factor of 1.5–2 and equal to 0.4–0.5%/year for women and 0.2–0.3%/year for men. Changes in BMC have been studied in detail in Asian population-based studies. Age-dependent BMC in persons older than 50 years can be evaluated as 0.6–0.9%/year in men and about 0.7–1.0%/year in women (Yao et al., 2001). Another study found higher rates of longitudinal BMC loss: 2%/year to 2.5%/year during menopausal transition and about 1.5%/year in early postmenopausal women (Ho et al. 2008). Because the rate of cortical bone resorption increases in women after menopause up to 6%/year (Figure 12), it is expected that the rate of bone apposition increases in this age interval up to 4–5%/year to compensate for accelerated bone loss, so that the decrease in vBMD and BMC does not exceed levels obtained by the other investigators. For men (Figure 12), the rate of cortical bone resorption increases after 55 years of age and reaches 3.5–4%/year. This indicates that the rate of bone apposition in men remains constant or slightly increases.

In summary, the following age-dependent changes in rates of cortical bone resorption were obtained; these studies can be used for modeling of bone-seeking elements in the human body for purposes of dosimetry and toxicology, and for modeling bone-turnover processes:

- The rate of endocortical resorption is independent of age in adult men less than 55-years old and in adult women before menopause. In this age period, the rates of cortical bone resorption are independent of gender and equal to 2.7%/year;
- In women after menopause, the rate of cortical bone resorption depends on time elapsed from the onset of menopause. The rate increases by a factor of two (up to 6%/year) compared to the period before menopause and remains at this level for 10–12 years; after that, there is a decline in the rate of cortical bone resorption to 5.0–5.5%/year;
- In men after 55 years of age there is a gradual age-dependent increase in the rate of cortical bone resorption up to 4%/year at ages >70 years; and
- Data obtained in this study indicate a significant increase in the rate of periosteal apposition in women after menopause, opposite to that seen in men of similar ages, to compensate for increased resorption.

Conclusion

Structural features of the cortical bone, such as significant surface area of bone minerals and low (compared to trabecular bone) blood-surface contact area, determine the ability of the cortical bone to accumulate xenobiotics and, in particular, bone-seeking contaminants. These contaminants include strontium and its anthropogenic radioactive isotopes. Release of these isotopes into the environment results in contamination of water and foodstuffs, and consequently, accumulation in human bones as was shown in this Chapter with an example of the Techa River contamination in Chelyabinsk Oblast, Russia. Ingested (or inhaled) contaminants can be absorbed on surfaces of hydroxyapatite crystals or incorporated in crystal lattice. The main mechanism for elimination of contaminants from the cortical bone is resorption by osteoclasts, which is overall lower than in trabecular bone. Significant accumulation of contaminants in cortical bone occurs during maximum skeletal growth in adolescence, or during the age of 12–15 years old. During this age period there is a significant increase in the skeletal mass, particularly in the cortical bone mass. Differences in accumulation of contaminants in adults and adolescents reach a factor of 4 and between adolescents and children of the first years of life such difference is about 5-fold. Contaminants accumulated in large amounts in the skeleton during adolescence can be detected in the cortical bone even decades afterwards.

The influence of pregnancy and lactation on accretion of trace elements in the maternal body had been insufficiently studied, because the main attention was usually paid to accumulation of minerals in the fetus and infant. Significant changes occur in maternal mineral turnover during pregnancy and lactation aimed at meeting the requirements of minerals by the fetus/infant. Such changes during pregnancy result in an increase of Ca and Sr gastrointestinal absorption; and Ca and Sr urinary excretion and bone-remodeling rate. Our study of about 100 women from the Techa Riverside villages shows an increase in accumulation of bone-seeking ^{90}Sr in the maternal skeleton during pregnancy and breastfeeding by a factor of 2 compared to non-pregnant, non-lactating women. Such an increase results from increased consumption of water and foodstuffs during pregnancy and lactation, as well as changes in the rates of bone-mineral turnover in the mother. Water and foodstuff consumption by women mostly increases at lactation (by 25%) than at pregnancy (by 5-10%). Our results indicate an increase in accumulation of toxic elements in pregnant/breastfeeding women compared to non-pregnant and non-lactating women.

Previously published data on cortical bone-resorption rates in adults are not complete. The results of our study based on the use of accidentally ingested ^{90}Sr as a tracer for bone metabolism showed that the rate of endocortical resorption is independent of age in adult men less than 55-years old and in adult women before menopause. In this age period, the rates of cortical bone resorption are independent of gender and equal to 2.7%/year. In older persons the cortical bone resorption increases. In men there is a continuous gradual increase after 55 years from 2.8 to 4.5%/year by the age of 75 years. In women, there is a doubled increase in the rate of cortical bone resorption after menopause of up to 6%/year; then the rate remains unchanged for 10–12 years with a subsequent gradual decline down to 5–5.5%/year. Comparison of the rate of cortical bone resorption in men and women older than 55 years showed that women expressed significantly higher levels of cortical resorption.

References

Abrams, S. A, O'Brien, K. O. & Stuff, J. E. (1996). Change in calcium kinetics associated with menarche. *J Clin Endocrino Metab*, *81*, 2017-2020.

Akleyev, A. V., Kossenko, M. M., Silkina, L. A., Degteva, M. O., Yachmenyov, V. A., Awa, A., Akiyama, M., Veremeeva, G. A., Vozilova, A. V. & Vyushkova, O. V. (1995). Health effects of radiation incidents in the Southern Urals. *Stem Cells.*, *13*(Supp. 11), 58-68.

Boecker, B., Hall, R., Inn, K, Lawerence, J., Ziemer. P., Eisele, G., Wachholz, B. & Burr, W. (1991). Current status of bioassay procedures to detect and quantify previous exposures to radioactive materials. *Health Phys*, *60* Suppl 1, 45–100.

Bougrov, N. G., Degteva, M. O. & Arshansky, S. M. (2008). Modernized whole-body counter SICH-9.1M for in vivo measurements of ^{90}Sr and ^{137}Cs body burden. Medicine of Extreme Situations (*Scientific and Practical Journal of Medical-Biological Agency of RF*), *4*(26), 78–86 (in Russian).

Bronner, F. & Abrams, S. A. (1998) Development and regulation of calcium metabolism in healthy girls. *J Nutr*, *128*, 1474-1480.

Bronner, F. (2000). Metals in bone: Aluminum, boron, cadmium, chromium, lead, silicon, and strontium. In: Bilezikian JP, Raisz LG, Rodan GA (eds) *Principles of bone biology*. 3rd edn. New York: Academic Press, 295-304.

Bronner, F., Aubert. J. -P., Richelle, L. J., Saville, P. D., Nicholas, J. A. & Cobb, J. R. (1963) Strontium and its relation to calcium metabolism. *J Clin Invest*, *42*, 1095-1104.

Cabrera, W. E., Schrooten, I., De Broe, M. E. & Haese, P. C. D. (1999). *Strontium and bone J Bone Miner Res*, *14*(5), 661-668.

Colnot, C. (2009). Skeletal cell fate decisions within periosteum and bone marrow during bone regeneration. *J Bone Mineral Res*, *24*, 274-282.

Cross, N. A., Hillman, L. S., Allen, S. H., Krause, G. F. & Vieira, N. E. (1995). Calcium homeostasis and bone metabolism during pregnancy, lactation, and postweaning: a longitudinal study. *Am J Clin Nutr*, *61*, 514-523.

Degteva, M. O. & Kozheurov, V. P. (1994). Age-dependent model for strontium retention in human bone. *Radiat Prot Dosim*, *53*, 229-234.

Degteva, M. O., Kozheurov, V. P., Tolstykh, E. I., Vorobiova, M. I., Anspaugh, L. R., Napier, B. A. & Kovtun, A. N. (2000). The Techa River Dosimetry System: Methods for the reconstruction of internal dose. *Health Phys*, *79*, 24-35.

Degteva, M. O., Shagina, N. B., Tolstykh, E. I., Bougrov, N. G., Zalyapin, V. I., Anspaugh, L. R. & Napier, B. A. (2007). An approach to reduction of uncertainties in internal doses reconstructed for the Techa River population. *Radiat Prot Dosimetry*, *127*, 480-485.

Degteva, M. O., Shagina, N. B., Vorobiova, M. I., Anspaugh, L. R. & Napier, B. A. (2011). Re-evaluation of waterborne releases of radioactive materials from “Mayak” Production Association into the Techa River in 1949–1951. *Health Phys*, *109*(2), 25-38.

Degteva, M. O., Vorobiova, M. I., Kozheurov, V. P., Tolstykh, E. I., Anspaugh, L. R. & Napier, B. A. (2000a). Dose reconstruction system for the exposed population living along the Techa River. *Health Phys*, *78*, 542-554.

Degteva, M. O., Vorobiova, M. I., Tolstykh, E. I., Shagina, N. B., Shishkina, E. A., Anspaugh, L. R., Napier, B. A. Bougrov, N. G., Shved, V. A. & Tokareva, E. E. (2006). Development of an improved dose reconstruction system for the Techa River population affected by the operation of the Mayak Production Association. *Radiat Res*, *166*, 255-270.

Dehos, R. & Schmier, H. (1985). Strontium-90 Content in Human Bone in West German Residents. In: Ed. Priest N.D. Metals in bone. Proceedings of a EULEP symposium on the deposition, retention and effects of radioactive and stable metals in bone and bone marrow tissues, October 11-13, 1984. Angers. France: MTP Press Limited Falcon House Lancaster, England. 87– 95.

DNC News: Strontium: A News Treatment for Osteoporosis. Danver Naturopatic Clinic. Information from the practices of Drs Scher J.J. and Bloom R.A.: http://www.denvernaturopathic.com/news/strontiumone.html; 2008, March. 10.

Donovan, B. T. & Van Der Weref Ten Bosch, J. J. (1965). *Physiology of puberty*. London: Edward Arnold, 19-24.

Driessens, F. C. M. & Verbeek, R. M. H. Eds. (1990). *Biominerals*. Roca Raton, Ann Arbor. Boston: CRP Press.

Engström, A. (1958). *Bone and Radiostrontium*. New York: Wiley.

Fell, T. P., Harrison, J. D. & Leggett, R. W. (1998). A model for the transfer of calcium and strontium to the fetus. *Radiat. Prot. Dosim*, *79*, 311– 316.

Heaney, R. P. & Skillman, T. G. (1971). Calcium metabolism in normal human pregnancy, *J Clin Endocrinol Metab*, *33*(4), 661– 670.

Ho, S. C., Chan, S. G., Yip, Y. B., Chan, C. S., Woo, J. L. & Sham, A. (2008). Change in bone mineral density and its determinants in pre- and perimenopausal Chinese women: The Hong Kong perimenopausal women osteoporosis study. *Osteoporos Int*, *19*, 1785–1796.

ICRP International Commission on Radiological Protection. (1975). Report of the Task Group on Reference Man. Publication 23. Oxford: Rergamon press.

ICRP International Commission on Radiological Protection. (1995). Basic Anatomical and physiological data for use in radiological protection: The skeleton. Publication 70. *Ann. ICRP*, *25*(2), Oxford: Rergamon press.

ICRP. (1992). International Commission on Radiological Protection, Age-dependent dose to members of the public from intake of radionuclides: Part 2: Ingestion dose coefficients. ICRP Publication 67. *Ann. ICRP, 23*(3/4). Oxford: Rergamon press.

Khosla, S., Melton, L. J. III, Atkinson, E. J., O'Fallon, W. M., Klee, G. G. & Riggs, B. L. (1998). Relationship of serum sex steroid levels and bone turnover markers with bone mineral density in men and women: A key role for bioavailable estrogen. *J Clin End Metab*, *83*, 2266–2274.

Kovacs, C. S. & Chada, E. -H. F. (2006) Calcium and bone disorders during pregnancy and lactation. *Endocrinol Metab Clin N Am*, *35*, 21-51.

Kozheurov, V. P. (1994). SICH-9.1—A unique whole-body counting system for measuring Sr-90 via bremsstrahlung: The main results from a long-term investigation of the Techa River population. *Sci Total Environ*, *14*, 37-48.

Kozheurov, V. P., Zalyapin, V. I., Shagina, N. B., Tokareva, E. E., Degteva, M. O., Tolstykh, E. I., Anspaugh, L. R. & Napier, B. A. (2002). Evaluation of uncertainties in the 90Sr-body-burdens obtained by whole-body count: Application of Bayes' Rule to derive detection limits by analysis of a posteriori data. *Appl Radiat Isot*, *57*, 525-535.

Krebs, N. F., Reidinger, C. J., Robertson, A. D. & Brenner, M. (1997). Bone mineral density changes during lactation: maternal dietary, and biochemical correlates. *Am J Clin Nutr*, *65*, 1738-1746.

Krestinina, L. Yu., Davis, R., Ostroumova, E. V., Epifanova, S. B., Degteva, M. O., Preston, D. L. & Akleyev, A. V. (2007). Solid cancer incidence and low-dose-rate radiation exposures in the Techa River Cohort: 1956–2002. *Intl J Epidemiol*, *36*, 1038-1046.

Krestinina, L. Yu., Preston, D. L., Ostroumova, E. V., Degteva, M. O., Ron, E., Vyushkova, O. V., Startsev, N. V., Kossenko, M. M. & Akleyev, A. V. (2005). Protracted radiation exposure and cancer mortality in the Techa River Cohort. *Radiat Res*, *164*, 602-611.

Krestinina, L., Preston, D. L., Davis, F. G., Epifanova, S., Ostroumova, E., Ron, E. & Akleyev, A. (2010) Leukemia incidence among people exposed to chronic radiation from the contaminated Techa River, 1953–2005. *Radiat Environ Biophys*, *49*, 195–201.

Lauretani, F., Bandinelli, S., Griswold, M. E., Maggio, M., Semba, R., Guralnik, J. M. & Ferrucci, L. (2008). Longitudinal changes in BMD and bone geometry in a population-based study. *J Bone Miner Res*, *23*, 400-408.

Leggett, R. W. (1992). A generic age specific biokinetic model for calcium-like elements. *Radiat Prot Dosim*, *41*, 183-198.

Lenihan, J. M. A., Loutit, J. F. & Martin, J. H. (Ed) (1967). Strontium metabolism. Proceedings of the international symposium on some aspects of strontium metabolism held at Chapelcross, *Glasgow and Strontian*, 5–7 May, 1966. London and New York: Academic Press.

Likhtarev, I. A., Dobroskok, I. A., Ilyin, L. A., Krasnoschekova, G. P., Likhtareva, T. M., Smirnov, B. I., Sobolev, E. P., Shamov, V. P. & Shapiro, E. L. (1975). A study of certain characteristics of strontium metabolism in a homogeneous group of human subjects. *Health Phys*, *28*, 49-60.

Lloyd, R. D., Mays, C. W. & Atherton, D. R. (1976). Distribution of injected 226Ra and 90Sr in the Beagle skeleton. *Health Phys*, *30*, 183–189.

Lorenzo, J., Horowitz, M. & Choi, Y. (2008). Osteoimmunology: Interactions of the bone and immune system. *Endocrinol Rev*, *29*, 403-440.

Lubashevsky, N. M. (1980). *Strontium metabolism in mammalian skeleton*. Moscow: Nauka (in Russian).

Manolagas, S. C. (1995). Role of cytokines in bone resorption. *Bone*, *17*, 63S-37S.

Manolagas, S. C. (2000). Birth and death of bone cells: Basic regulatory mechanism and implications for the pathogenesis and treatment of osteoporosis. *Endocrinol Rev*, *21*, 115-137.

Marshall, J. H. (1972). *Alkaline earth metabolism in adult man: a condensation*. In: Proceedings of the second International conference on strontium metabolism. Springfield, VA: National Technical Information Service; CONF-720818, 207-228.

Papworth, D. G. & Vennart, J. (1984). The uptake and turnover of Sr-90 in the human skeleton. *Phys Med Biol*, *29*, 1045–1061.

Parfitt, A. M. (2006). Misconceptions V—Activation of osteoclasts is the first step in the bone remodeling cycle. *Bone*, *39*, 1170-1172.

Pitkin, R. M. (1985). Calcium metabolism in pregnancy and the perinatal period: A review *Am. J. Obstet Gynecol*, *151*, 99– 109.

Recker, R., Lappe, J., Davies, K. M. & Heaney, R. (2004). Bone remodeling increases substantially in the years after menopause and remaind increased in older osteoporosis patients. *J Bone Mineral Res*, *19*, 1628-1633.

Riggs, B. L. & Melton, L. J. III (Ed) (1996). Osteoporosis. *Etiology, diagnosis and management*. Second Edition, Philadelphia: Lippincott.

Riggs, B. L., Melton, L. J., Robb, R. A., Camp, J. J., Atkinson, E. J., McDaniel, L., Amin, S., Rouleau, P. A. & Khosla, S. (2008). A population-based assessment of rates of bone loss at multiple skeletal sites: Evidence for substantial trabecular bone loss in young adult women and men. *J Bone Miner Res*, *23*, 205–214.

Ritchie, L. D., Fung, E. B., Halloran, B. P., Turnlund, J. R., Van Loan M. D., Cann, C. E. & King, J. C. (1998). A longitudinal study of calcium homeostasis during human pregnancy and lactation and after resumption of menses. *Am J Clin Nutr*, *67*, 693-701.

Schindeler, A., Liu, R. & Little, D. G. (2009). The contribution of different cell lineages to bone repair: Exploring a role for muscle stem cells. *Differentiation*, *77*, 12-18.

Shagina, N. B., Tolstykh, E. I. & Degteva, M. O. (2003b). Improvements in the biokinetic model for strontium with allowance for age and gender differences in bone mineral metabolism. *Radiat Prot Dosim*, *105*. 619-622.

Shagina, N. B., Tolstykh, E. I., Degteva, M. O., Anspaugh, L. R. & Napier, B. A. (2012). Cortical bone resorption rate in elderly persons: estimates from long-term in vivo measurements of (90)Sr in the skeleton. *Arch Gerontol Geriatr*, *54*(3), e411-418.

Shagina, N. B., Tolstykh, E. I., Zalyapin, V. I., Kozheurov, V. P., Tokareva, E. E., Anspaugh, L. R. & Napier, B. A. (2003a). Evaluation of age and gender dependences of the rate of strontium elimination 25–45 years after intake: Analysis of data from residents living along the Techa River. *Radiat Res*, *159*, 239-246.

Simmons, K. & Greuilich, W. W. (1942). Monarchal age and the height, weight and skeletal age of girls age 7 to 17 years. *J. Pediat*, *22*, 518– 548.

Staub, J. F., Foos, E., Courtin, B., Jochemsen, R. & Perault-Staub, A. M. (2003). A nonlinear compartmental model of Sr metabolism. II. Its physiological relevance for Ca metabolism. *Am J Physiol Regul Integr Comp Physiol*, *284*, 835–852.

Staub, J._F., Foos, E., Courtin, B., Jochemsen, R. & Perault-Staub, A. M. (2003). A nonlinear compartmental model of Sr metabolism. I. Non-steady-state kinetics and model building. *Am J Physiol Regul Integr Comp Physiol*, *284*(3), 819–834.

Sveshnikov, A. A. & Kuznetsov, A. P. (1989). Age-related changes in bone mineral contents in lower extremities of healthy humans. *Human Physiol*, *15*, 148–153 (in Russian).

Tandon, L., Iyengar, G. V. & Parr, R. M. (1998). A Review of Radiologically important Trace Elements in Human Bones. *Appl Radiat Isot*, *49*(8), 903–910.

Teitelbaum, S. L. (2000). Bone resorption by osteoclasts. *Science*, *289*, 1504-1508.

Teitelbaum, S. L. (2007). Osteoclasts: What do they do and how do they do it? *Am J Pathol*, *170*, 427-435.

Tolstykh, E. I., Degteva, M. O., Kozheurov, V. P., Shishkina, E. A., Romanyukha, A. A., Wieser, A. & Jacob, P. (2000). Strontium metabolism in teeth and enamel dose assessment: Analysis of the Techa River data. *Radiat Environ Biophys*, *39*, 161–171.

Tolstykh, E. I., Degteva, M. O., Peremyslova, L. M., Shagina, N. B., Shishkina, E. A., Krivoshchapov, V. A., Anspaugh, L. R. & Napier, B. A. (2011b). Reconstruction of long-lived radionuclide intakes for Techa riverside residents: strontium-90. *Health Phys*, *101*(1), 28-47.

Tolstykh, E. I., Kozheurov, V. P., Degteva, M. O. & Burmistrov, D. S. (1998). Strontium Transfer from Maternal Skeleton to the Fetus Estimated on the Basis of the Techa River Data. *Rad Prot Dosim*, *79*(1–4), 307–310.

Tolstykh, E. I., Kozheurov, V. P., Vyushkova, O. V. & Degteva, M. O. (1997). Analysis of strontium metabolism in humans on the basis of the Techa River Data. *Radiat Environ Biophys*, *36*, 25-29.

Tolstykh, E. I., Shagina, N. B., Degteva, M. O., Anspaugh, L. R. & Napier, B. A. (2011a). Does the rate of cortical bone resorption change due to 90Sr-re into consideration, is still lacking. In the present work, an nicro-CT finite element model at bone tissue level coupled to a w was developed in order to simulate the fracture of human bone specimens under quasi-static compressive load and apparent stress and strain at fracture and the final fracture the specimens. The element deletion technique was applied in imulate the progressive fracturing process of bone tissue. To esh-dependence that generally affects the damage propagation arization technique was applied in the current work. The model validated with experimental results performed on seven specimens. It was then applied to assess the apparent stress and fracture of twenty-three human trabecular bone specimens from different bone sites. The simulation results were with reported experimental results to investigate the accuracy ro-FE model in predicting the apparent failure properties of bone. It was found that the predicted failure properties agreed vith the experimental ones.

abecular bone fracture, Micro-finite element damage modeling, nitiation and propagation, Apparent stress and strain tissue

1. Introduction

of elderly people with osteoporosis are susceptible to traumatic esult of external impact. At the macroscopic scale, bone is ortical bone and trabecular bone. Trabecular bone is a major ological tissue in human femurs and vertebrae. It is involved in ctures and is the primary site for the insertion of orthopedic parent mechanical properties are of great clinical and research proper treatment to strengthen the bone in order to prevent such sign better implants.

the apparent fracture behavior of trabecular bone provides cture mechanisms in bones and enables the effects of aging, ig treatments on bone strength to be assessed. In such cases, complex process that strongly depends on several factors volume fraction, microrchitecture, the mechanical properties of and the applied loads.

radiation exposure? Analysis of data from Techa Riverside residents. *Radiat Environ Biophys*, *50*(3), 417-430.

Tolstykh, E. I., Shagina, N. B., Peremyslova, L. M., Degteva, M. O., Phipps, A. W., Harrison, J. D. & Fell, T. P. (2008). Reconstruction of (90)Sr intake for breast-fed infants in the Techa riverside settlements. *Radiat Environ Biophys*, *47*(3), 349-357.

Turner, C. H. (1998). Three rules for bone adaptation to mechanical stimuli. *Bone*, *23*, 399-407

UNSCEAR 2000. (2000). Sources and effects of ionizing radiation. Report to the General Assembly, with Scientific Annexes United Nations. New York.

Vorobiova, M. I., Degteva, M. O., Burmistrov, D. S., Safronova, N. G., Kozheurov, V. P., Anspaugh, L. R. & Napier, B. A. (1999). Review of Historical Monitoring Data on the Techa River Contamination. *Health Phys*, *76*(6), 605-618.

Wasserman, R. H. (1998). Strontium as a tracer for calcium in biological and clinical research. *Clin Chem*, *44*(3), 437-439.

Watchman, C. J., Bourke, V. A., Lyon, J. R., Knowlton, A. E., Butler, S. L., Grier, D. D., Wingard, J. R., Braylan, R. C. & Bolch, W. E. (2007). Spatial distribution of blood vessels and CD34+ hematopoietic stem and progenitor cells within the marrow cavities of human cancellous bone. *J Nucl Med*, *48*, 645-654.

Yao, W. J., Wu, C. H., Wang, S. T., Chang, C. J., Chiu, N. T. & Yu, C. Y. (2001). Differential changes in regional bone mineral density in healthy Chinese: age-related and sex-dependent. *Calcif Tissue Intl*, *68*, 330-336.

Zaichick, S. & Zaichick, V. (2010). Human bone as a biological material for environmental monitoring. *Int. J. Environment and Health*, *4*(2/3), 278–292.

In: Trabecular and Cortical Bone
Editor: V. Subramanium, M. Bhatnagar © 2

Evaluation of Ap
and Strain at the
Human Trabecul
Quasi-Static Com
Using Micro
Element Si

Ridha H
Prisme Institute/MMH, O

Abst

Micro-finite element (micro-FE)
a standard tool for the assessment of
trabecular bone. Most micro-CT FE
focused on linear and non-linear
properties. However, prediction of
bone, including the damage process
taking crack initiation and its prog

* Prisme Institute/MMH, 8, Rue Léonard de Vin
(0)2-38-49-40-55, Mail : ridha.hambli@uni

bone fai
isotropic
damage
trabecula
predict t
pattern o
order to
prevent r
rate, regu
was firs
trabecula
strain at
harvested
compared
of the m
trabecular
very well

Keywords: T
Damage
propertie

The bone
fracture as a
composed of
load-bearing b
bone organ fr
implants. Its a
interest to find
fractures and d
Modeling
insight into fr
disease, and d
fracture is a
including bone
the bone tissue

Micro-CT finite element models (μFE) are used extensively to study the mechanical properties of trabecular bone, both at the continuum level and at the microstructural level.

Non-linear μFE were used to predict trabecular bone yield properties (van Rietbergen et al., 1998; Niebur et al., 2000; Stölken and Kinney, 2003; Bayraktar and Keaveny 2004; Kinney and Stölken, 2005; Verhulp et al., 2008). However, (i) simulation of the initiation and subsequent progressive propagation of cracks within the trabecula that take place during the fracturing process was not considered and (ii) validation of the proposed FE models with experimental results is still lacking. van Rietbergen et al. (1998) and Verhulp et al. (2008) reported that simple non-linear models could predict the ultimate stress but not the ultimate strain of the tissue. More advanced material models were used in recent studies. Bevill et al. (2006) reported that inclusion of large-deformation effects in high-resolution finite element models for trabecular bone improved the prediction of strength versus experiment. Several authors investigated the fracture of cortical and trabecular bone based on fracture mechanics concepts (Ural et al., 2007; Yang et al., 2006; Nalla et al., 2005; Vashishth, 2004; Malik et al., 2003) and the extended FE method (X-FEM) (Budyn et al., 2008, Abdel-Wahab and Silberschmidt, 2011). Tomar (2008) developed a cohesive FE model to simulate the fracture process that considered the loading rates. The author performed FE analyses on 2D sections but did not take the 3D connectivity of trabecular bone into account. Nevertheless, cohesive FE methods are restricted to the problem of idealized planar cracks with a cohesive zone (Yang et al., 2006) and these methods are based on a limited number of simple uncoupled crack initiation criteria (principal stress or strain) which are unable to describe complex bone damage and fracture behavior or to distinguish between tension and compression (Keaveny et al., 1994; Currey, 2002; Bayraktar et al., 2004). Recently, Harrison et al. (2012) a 3D FE model to simulate complete fracture of ovine vertebral trabecular bone samples incorporating damage growth and fracture criteria. The proposed model incorporated a cohesive parameter representing the damage and fracture variables and the cracks propagation was simulated by the element deletion technique (the broken FE of the mesh is removed when the critical strain is reached).

Despite the high clinical relevance of trabecular bone fractures, there is a lack of validated FE models describing the complete fracture behavior of trabecular bone specimens. A quantitative assessment of trabecular level conditions associated with initiation of microdamage till total fracture may provide insight into the improvement of fracture risk assessment methods and the development of therapeutic strategies for the treatment of skeletal fragility diseases such as osteoporosis.

An alternative method to simulate in a simple, accurate and realistic way crack initiation and propagation under an applied excessive load is continuum damage mechanics (CDM) (Lemaitre, 1985; Hambli et al., 2012a,b; Hambli, 2012). With the application of CDM, one can simulate the initiation and progressive growth of cracks based on specific experimental brittle to quasi-brittle damage laws (Reilly and Burstein, 1975; Currey, 2002; Kaneko et al., 2003; Bayraktar et al., 2004).

In the present work, an isotropic μFE model coupled to a damage law was developed in order to simulate fracture of human trabecular bone specimens under quasi-static compressive load and predict the apparent stress and strain at fracture and the final fracture pattern of the specimens. The element deletion technique was applied in order to simulate the progressive fracturing process of bone. To prevent mesh-dependence that generally affects the damage propagation rate, regularization technique was applied in the current work. The model was first validated with experimental results performed on 7 specimens. It was then applied to assess the apparent stress and strain at fracture of 23 human trabecular bone specimens. The simulation results were compared with reported experimental results to investigate the accuracy of the μFE model in predicting the apparent failure properties of trabecular bone. It was found that the predicted failure properties agreed very well with the experimental ones.

2. Material and Methods

2.1. FE Simulation of Bone Fracture Initiation and Propagation under Quasi-Static Load

Recently, several anisotropic viscoelastic behavior laws coupled to damage for macroscopic bone at the organ level have been developed by various authors (Natali et al., 2008; Garcia et al., 2009; Hosseini et al., 2012). However, the main limitation of application of these models is that these sophisticated models need a large number of parameters (anisotropy, viscosity and damage) that are not available in the published literature and are very complicated to measure experimentally.

In general, at a low load rate (quasi-static regime), the bone behaves as a material with a non-linear behavior till complete fracture (Reilly and Burstein, radiation exposure? Analysis of data from Techa Riverside residents. *Radiat Environ Biophys*, *50*(3), 417-430.

Tolstykh, E. I., Shagina, N. B., Peremyslova, L. M., Degteva, M. O., Phipps, A. W., Harrison, J. D. & Fell, T. P. (2008). Reconstruction of (90)Sr intake for breast-fed infants in the Techa riverside settlements. *Radiat Environ Biophys*, *47*(3), 349-357.

Turner, C. H. (1998). Three rules for bone adaptation to mechanical stimuli. *Bone*, *23*, 399-407

UNSCEAR 2000. (2000). Sources and effects of ionizing radiation. Report to the General Assembly, with Scientific Annexes United Nations. New York.

Vorobiova, M. I., Degteva, M. O., Burmistrov, D. S., Safronova, N. G., Kozheurov, V. P., Anspaugh, L. R. & Napier, B. A. (1999). Review of Historical Monitoring Data on the Techa River Contamination. *Health Phys*, *76*(6), 605-618.

Wasserman, R. H. (1998). Strontium as a tracer for calcium in biological and clinical research. *Clin Chem*, *44*(3), 437-439.

Watchman, C. J., Bourke, V. A., Lyon, J. R., Knowlton, A. E., Butler, S. L., Grier, D. D., Wingard, J. R., Braylan, R. C. & Bolch, W. E. (2007). Spatial distribution of blood vessels and CD34+ hematopoietic stem and progenitor cells within the marrow cavities of human cancellous bone. *J Nucl Med*, *48*, 645-654.

Yao, W. J., Wu, C. H., Wang, S. T., Chang, C. J., Chiu, N. T. & Yu, C. Y. (2001). Differential changes in regional bone mineral density in healthy Chinese: age-related and sex-dependent. *Calcif Tissue Intl*, *68*, 330-336.

Zaichick, S. & Zaichick, V. (2010). Human bone as a biological material for environmental monitoring. *Int. J. Environment and Health*, *4*(2/3), 278–292.

In: Trabecular and Cortical Bone
ISBN: 978-1-62808-766-6
Editor: V. Subramanium, M. Bhatnagar © 2014 Nova Science Publishers, Inc.

Chapter III

Evaluation of Apparent Stress and Strain at the Fracture of the Human Trabecular Bone under Quasi-Static Compression Load Using Micro-CT Finite Element Simulation

Ridha Hambli [*]
Prisme Institute/MMH, Orléans cedex 2, France

Abstract

Micro-finite element (micro-FE) analysis is considered nowadays as a standard tool for the assessment of the mechanical properties of human trabecular bone. Most micro-CT FE analysis of this type of bone has focused on linear and non-linear analysis to evaluate bone failure properties. However, prediction of the failure properties of trabecular bone, including the damage process under monotonic excessive load and taking crack initiation and its progressive propagation until complete

[*] Prisme Institute/MMH, 8, Rue Léonard de Vinci, 45072 Orléans cedex 2, France, Phone : +33 (0)2-38-49-40-55, Mail : ridha.hambli@univ-orleans.fr.

bone failure into consideration, is still lacking. In the present work, an isotropic micro-CT finite element model at bone tissue level coupled to a damage law was developed in order to simulate the fracture of human trabecular bone specimens under quasi-static compressive load and predict the apparent stress and strain at fracture and the final fracture pattern of the specimens. The element deletion technique was applied in order to simulate the progressive fracturing process of bone tissue. To prevent mesh-dependence that generally affects the damage propagation rate, regularization technique was applied in the current work. The model was first validated with experimental results performed on seven trabecular specimens. It was then applied to assess the apparent stress and strain at fracture of twenty-three human trabecular bone specimens harvested from different bone sites. The simulation results were compared with reported experimental results to investigate the accuracy of the micro-FE model in predicting the apparent failure properties of trabecular bone. It was found that the predicted failure properties agreed very well with the experimental ones.

Keywords: Trabecular bone fracture, Micro-finite element damage modeling, Damage initiation and propagation, Apparent stress and strain tissue properties

1. Introduction

The bones of elderly people with osteoporosis are susceptible to traumatic fracture as a result of external impact. At the macroscopic scale, bone is composed of cortical bone and trabecular bone. Trabecular bone is a major load-bearing biological tissue in human femurs and vertebrae. It is involved in bone organ fractures and is the primary site for the insertion of orthopedic implants. Its apparent mechanical properties are of great clinical and research interest to find proper treatment to strengthen the bone in order to prevent such fractures and design better implants.

Modeling the apparent fracture behavior of trabecular bone provides insight into fracture mechanisms in bones and enables the effects of aging, disease, and drug treatments on bone strength to be assessed. In such cases, fracture is a complex process that strongly depends on several factors including bone volume fraction, microrchitecture, the mechanical properties of the bone tissue, and the applied loads.

Micro-CT finite element models (μFE) are used extensively to study the mechanical properties of trabecular bone, both at the continuum level and at the microstructural level.

Non-linear μFE were used to predict trabecular bone yield properties (van Rietbergen et al., 1998; Niebur et al., 2000; Stölken and Kinney, 2003; Bayraktar and Keaveny 2004; Kinney and Stölken, 2005; Verhulp et al., 2008). However, (i) simulation of the initiation and subsequent progressive propagation of cracks within the trabecula that take place during the fracturing process was not considered and (ii) validation of the proposed FE models with experimental results is still lacking. van Rietbergen et al. (1998) and Verhulp et al. (2008) reported that simple non-linear models could predict the ultimate stress but not the ultimate strain of the tissue. More advanced material models were used in recent studies. Bevill et al. (2006) reported that inclusion of large-deformation effects in high-resolution finite element models for trabecular bone improved the prediction of strength versus experiment. Several authors investigated the fracture of cortical and trabecular bone based on fracture mechanics concepts (Ural et al., 2007; Yang et al., 2006; Nalla et al., 2005; Vashishth, 2004; Malik et al., 2003) and the extended FE method (X-FEM) (Budyn et al., 2008, Abdel-Wahab and Silberschmidt, 2011). Tomar (2008) developed a cohesive FE model to simulate the fracture process that considered the loading rates. The author performed FE analyses on 2D sections but did not take the 3D connectivity of trabecular bone into account. Nevertheless, cohesive FE methods are restricted to the problem of idealized planar cracks with a cohesive zone (Yang et al., 2006) and these methods are based on a limited number of simple uncoupled crack initiation criteria (principal stress or strain) which are unable to describe complex bone damage and fracture behavior or to distinguish between tension and compression (Keaveny et al., 1994; Currey, 2002; Bayraktar et al., 2004). Recently, Harrison et al. (2012) a 3D FE model to simulate complete fracture of ovine vertebral trabecular bone samples incorporating damage growth and fracture criteria. The proposed model incorporated a cohesive parameter representing the damage and fracture variables and the cracks propagation was simulated by the element deletion technique (the broken FE of the mesh is removed when the critical strain is reached).

Despite the high clinical relevance of trabecular bone fractures, there is a lack of validated FE models describing the complete fracture behavior of trabecular bone specimens. A quantitative assessment of trabecular level conditions associated with initiation of microdamage till total fracture may provide insight into the improvement of fracture risk assessment methods and the development of therapeutic strategies for the treatment of skeletal fragility diseases such as osteoporosis.

An alternative method to simulate in a simple, accurate and realistic way crack initiation and propagation under an applied excessive load is continuum damage mechanics (CDM) (Lemaitre, 1985; Hambli et al., 2012a,b; Hambli, 2012). With the application of CDM, one can simulate the initiation and progressive growth of cracks based on specific experimental brittle to quasi-brittle damage laws (Reilly and Burstein, 1975; Currey, 2002; Kaneko et al., 2003; Bayraktar et al., 2004).

In the present work, an isotropic μFE model coupled to a damage law was developed in order to simulate fracture of human trabecular bone specimens under quasi-static compressive load and predict the apparent stress and strain at fracture and the final fracture pattern of the specimens. The element deletion technique was applied in order to simulate the progressive fracturing process of bone. To prevent mesh-dependence that generally affects the damage propagation rate, regularization technique was applied in the current work. The model was first validated with experimental results performed on 7 specimens. It was then applied to assess the apparent stress and strain at fracture of 23 human trabecular bone specimens. The simulation results were compared with reported experimental results to investigate the accuracy of the μFE model in predicting the apparent failure properties of trabecular bone. It was found that the predicted failure properties agreed very well with the experimental ones.

2. Material and Methods

2.1. FE Simulation of Bone Fracture Initiation and Propagation under Quasi-Static Load

Recently, several anisotropic viscoelastic behavior laws coupled to damage for macroscopic bone at the organ level have been developed by various authors (Natali et al., 2008; Garcia et al., 2009; Hosseini et al., 2012). However, the main limitation of application of these models is that these sophisticated models need a large number of parameters (anisotropy, viscosity and damage) that are not available in the published literature and are very complicated to measure experimentally.

In general, at a low load rate (quasi-static regime), the bone behaves as a material with a non-linear behavior till complete fracture (Reilly and Burstein,

1975; Currey, 1990; Kotha and Guzelsu, 2003; Keaveny et al., 1994, 1999, 2001). In this case, behavior laws coupled to damage are well suited to represent the non-linear behavior before complete fracture observed during quasi-static bone specimen fracture experiments (Nagaraja et al., 2005; Parsamian, 2002).

The damage behavior of trabecular bone corresponds to the generation of microcracks (Vashishth et al., 2000), which leads to a reduction in stiffness, energy dissipation, and permanent strains, and may result in bone failure (Keaveny et al., 1999). In addition, a certain amount of deformation remains at zero stress, which cannot be described by standard elastic damage models. These permanent deformations can be taken into account by adding inelastic terms to the constitutive model, but since the aim of the proposed FE model is to simulate the fracture behavior of a monotonic single applied loading till complete fracture (without unloading), this extension will not be considered here for simplicity's sake.

Although bone tissue has been shown to behave as a viscoelastic inelastic damageable material for a large range of strain rates (Fondrk et al., 1988, 1999), rate-dependent effects have a moderate impact on physiological strain rates as they occur during normal daily activities (~ 1 Hz) (Fondrk et al., 1988). Hence, in the quasi-static load regime (low strain rate), viscous, thermal or other non-mechanical effects are not taken into account and strains and rotations are assumed to be small.

Trabecular bone tissue is anisotropic and inhomogeneous due to its multiscale hierarchical structure (Bevill et al., 2009). It has been reported that the use of trabecular bone tissue isotropic properties may be useful for bone mechanical strength analyses (Kabel et al., 1999). Therefore, in the quasi-static regime, the isotropic stress-strain relation of elasticity based damage mechanics is expressed by (Lemaitre, 1985; Hambli et al., 2012a,b):

$$\sigma_{ij} = (1 - D)C_{ijkl}\varepsilon_{kl} \quad (1)$$

where D denotes the damage variable, σ_{ij} the stress components, ε_{kl} the strains and C_{ijkl} are the elasticity tensor components.

The behavior of bone subjected to mechanical solicitations is characterized by diffuse microcracking that later localizes in relatively narrow zones, referred to as the fracture process zones. As a result, a gradual decrease in deformation resistance is usually observed instead of the sudden loss of strength observed in perfectly brittle fracture. Relation (1) shows that the damage variable acts as a stiffness reduction factor. For increasing damage, the effective stiffness moduli $(1-D)C_{ijkl}$ decrease.

Bone tissue damage in the form of numerous microcracks can be detected using different techniques including stress whitening effect, microscopy and staining techniques. In a recent study (Jungmann et al., 2011), a strategy for the direct measurement of local damage initiation and propagation in single trabeculae exploiting the stress whitening effect found in bone (Thurner et al., 2007) were applied. The authors suggested that damage initiation in trabecular bone is asymmetric in tension and compression, with damage/failure strongly depending on strain amplitude. It have been reported by other studies that the failure process of bone is controlled by strain (Nalla et al., 2003; Bayraktar and Keaveny, 2004; Keaveny et al., 1994, 1999, 2001, Niebur et al., 2000; Nagaraja et al., 2005; Wolfram et al., 2011). The amount of micro-damage in an experimental specimen has been positively correlated with applied post-yield strain (Arthur Moore and Gibson, 2002; Huang et al., 2008).

Nagaraja et al. (2005) used micro-CT and sequential fluorescent staining to assess local trabecular bone damage under varying levels of uniaxial compression. The authors performed incremental compression loads under controlled strain levels on fifteen fluorescent stained cylindrical specimens of bovine bone. The results showed significant positive correlations between stress and strain parameters and considerable microdamage within an individual trabecula observed by micro-CT imaging.

Considering the experimental results of Wolfram et al. (2011) performed on 251 cylindrical trabecular bone samples obtained from human vertebrae (T1-L3), an experimentally fitted damage law can be expressed in the general form:

$$\begin{cases} D = 0 & ; \varepsilon_{eq} \le \varepsilon_y \\ D = D_c\, \varepsilon_{eq}^n & ; \varepsilon_y < \varepsilon_{eq} < \varepsilon_f \\ D = D_c & ; \varepsilon_{eq} \ge \varepsilon_f \end{cases} \tag{2}$$

ε_{eq} , D_c, n, ε_y and ε_f are respectively the equivalent strain, the critical damage at fracture, the damage exponent, yield strain (damage strain threshold when damage starts) and the strain at fracture.

Numerous studies showed that damage threshold strains and stresses of trabecular and cortical bone tissue are different in tension and compression (Reilly and Burstein, 1975; Currey, 1990; Kotha and Guzelsu, 2003; Keaveny et al., 1994, 1999, 2001; Wolfram et al., 2011). Therefore, the role of pressure (hydrostatic stress) in the yield behavior of trabecular bone can be represented as different values of ultimate strains in tension and compression as follow:

$$\varepsilon_f = \varepsilon_f^T \text{ and } D_c = D_c^T \text{ in tension} \tag{3-a}$$

$$\varepsilon_f = \varepsilon_f^C \text{ and } D_c = D_c^C \text{ in compression and shearing} \tag{3-b}$$

ε_f^T and ε_f^C are the bone tissue tensile and compressive strain at fracture respectively

D_c^T and D_c^C are critical damage values at fracture in tension and compression.

2.1.1. Simulation of Crack Propagation Based an Element Deletion Method

Trabecular bone tissue fracture occurs in two stages: (i) damage initiation and propagation which induces the decrease of the elastic bone tissue properties and (ii) fracture of the trabeculae generating a complete separation of the damaged trabeculae which induces to the net decrease of the internal stresses/ forces transfert capacity by the broken trabeculae. Explicit FE Modeling of cracks propagation till complete fracture is a very complicated task. The FE mesh must be refined to a very small size to be able to represent the crack band and the mechanisms related to the crack propagation such as spatial variation of the crack band, the specific behavior of the ahead and behind the crack tip and the difference behavior in tension/ compression/shearing (Lemaitre, 1985; Nalla et al., 2003; Hambli et al., 2012).

In order to simulate crack initiation and propagation within the FE mesh, the element deletion method was applied (Hambli et al., 2012a,b). When the damage parameter reaches its critical value inside an element, its stiffness matrix is set to zero, leading to the redistribution of the stress state in the vicinity of the completely damaged zone (crack initiation). Once a crack is initiated, the propagation direction is simulated by the propagation of the broken elements of the mesh. The technique developed here distinguishes between tension, shearing and compression loading modes. At tissue level, the local critical damage value was set to (Hambli et al., 2012a,b).

In tension: $D_c^T = 0.95$.

In shearing and compression $D_c^C = 0.5$

A similar approach was applied by MacNeil and Boyd (2008) to develop a non-linear FE model to predict human distal radius strength. A yield criterion, based on maximum compressive and tensile yield strains, was incorporated into the constitutive model. When the tissue maximum principal tensile or compressive strain exceeded the yield values, the elastic properties of the tissue were reduced by 95%.

Recently, Wang et al. (2008), Hambli et al. (2012a,b) and Hambli (2012) successfully applied the element deletion technique to simulate crack initiation and propagation in human proximal femurs.

2.1.2. Regularization of Crack Propagation Dependence on Mesh Size

In the constitutive law of conventional CDM there is no intrinsic material characteristic length, which means that crack propagation results are dependent on the mesh size (Saanouni et al., 1989; Murakami and Liu, 1995; de Vree et al., 1995).

When the mesh is refined during damage FE analysis, crack growth prediction becomes more accurate in that the displacement gradients which describe the discontinuities become stronger. Consequently, the predicted critical damage at fracture occurs faster, leading to faster crack propagation (Abaqus, 2012). Therefore, when conducting an FE simulation of crack propagation, the physical crack length (L_{frx}) must be considered in relation with the local FE mesh characteristic length (L_{FE}). It has been reported that such an approach significantly minimizes the crack propagation dependence on the mesh size (Saanouni et al., 1989; Murakami and Liu, 1995; de Vree et al., 1995). A simple alternative method to ensure the objectivity of the numerical model in relation with the physical cracking process and the mesh dependence problem consists in weighting (linear form) the strain at fracture as a function of a characteristic FE length (L_{FE}) and the crack length (L_{frx}) in the form (Abaqus, 2012; Hambli, 2012):

$$\varepsilon_f^T = \varepsilon_{f-true}^T \left(\frac{L_{frx}}{L_{FE}} \right) \tag{4-a}$$

$$\varepsilon_f^C = \varepsilon_{f-true}^C \left(\frac{L_{frx}}{L_{FE}} \right) \tag{4-b}$$

$$L_{FE} = \left(V_{elem}\right)^{\frac{1}{3}} \tag{5}$$

where ε_{f-true}^T and ε_{f-true}^C denote respectively, the true bone material measured strain at fracture in tension and compression.

V_{elem} denotes the volume of the FE.

Average crack lengths found in bones are typically 100 microns (Taylor and Lee, 2003). The characteristic length L_{frx} was therefore set to ($L_{frx} = 0.1\ mm$).

In addition, the characteristic length calculation is based only on the element geometry. As a result, some level of mesh sensitivity remains related to the cracking direction. Therefore, elements with an aspect ratio of one were selected in the current work. The proposed mesh-dependence regularization method is recommended by the Abaqus code (Abaqus, 2012) for crack propagation problems to ensure non-dependence of the fracture results on the finite element mesh.

To investigate the potential of the mesh regularization technique, one experimentally tested specimen (among the seven used for the validation) was meshed with two different meshes (coarse and fine meshes) and simulated till complete fracture in compression.

2.2. Trabecular Bone Specimens

Cancellous porous bone has a complex architecture that varies substantially between anatomic sites and across individuals. Different trabecular samples were therefore used to capture different proximal femur morphologies. In the current study twenty-three cylindrical cores (7.04 mm in diameter and 5.5 to 10 mm long) from fresh-frozen elderly human proximal femur (n=12) and greater trochanter (n=11) trabecular bone were harvested from twelve males and females (50-94 years old). All the specimens were machined such that the main trabecular orientation was aligned with the axis of the core (Morgan and Keaveny, 2001). None of the donors had a history of metabolic bone disease or cancer. The 23 specimens were scanned using micro-computed tomography using the Skyscan 1072 system at an approximate spatial resolution of 20 μm Descriptive statistics of morphometric properties are given in Table 1.

Table 1. Descriptive statistics for morphometric analysis performed on 23 specimens in different human femur anatomic sites (femoral neck=12, greater trochanter=11) and for different gender (male=12, female=11)

	Age	BV/TV (%)	BS/TV (mm-1)	BS/BV (mm-1)	SMI	DA	Tb.Th (mm)	Tb.N (mm-1)	Tb.Sp (mm)	Tb.Pf (mm-1)	Conn. D (mm-3)
Mean	79,818	23,871	3,651	16,105	0,685	1,798	0,224	1,060	0,746	0,412	0,720
SD	8,000	3,735	0,325	2,124	0,198	0,141	0,031	0,095	0,065	0,364	0,278
Min.	50,000	12,612	2,803	9,480	0,023	1,338	0,157	0,802	0,557	1,282	0,347
Max.	98,000	35,019	4,753	22,227	1,214	2,181	0,341	1,372	0,946	3,465	2,015
CV (%)	10,023	15,645	8,900	13,192	28,946	7,814	13,669	8,964	8,703	88,358	38,618

BV/TV=bone volume fraction, BS/TV= bone surface per total volume, BV/BV= bone surface area to bone volume, SMI= bone structure model index, DA=bone degree of anisotropy, Tb.Th=trabecular bone thickness, Tb.N= trabecular bone number, Tb.Sp=trabecular bone spacing, Tb.Pf= trabecular bone pattern factor and Conn.D=connectivity density.

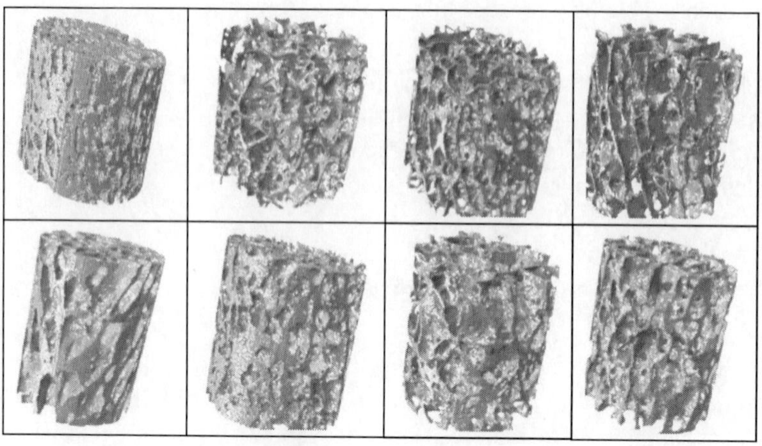

Figure 1. Eight examples of harvested trabecular cores from different anatomic sites of different human femurs.

Voxel FE at 20 μm voxel sizes were generated to simulate the fracture behavior of the trabecular samples under compressive load and predict the yield strength of each specimen. Some representative images from different anatomic sites are shown in Figure 1.

2.3. Validation Experiments

Compression experiments on seven specimens were conducted using a hydraulic material testing machine (Instron, model 4411) (Fig. 2-a). The quasi static compression load was measured with a commercial load cell applied at a constant velocity rate of 0.33 mm/sec. Axial strain was measured with an extensometer at room temperature. Testing was conducted at room temperature (approx. 20°C). Specimens were always kept wet with a saline solution drip at 37°C.

The force displacement curves were measured for each test and the ultimate apparent stress (σ_u) and strain (ε_u) were calculated by:

$$\sigma_u = \frac{F_u}{A} \tag{6}$$

and

$$\varepsilon_u = \frac{\Delta l_u}{l_0} \tag{7}$$

where A, F_u, Δl_u and l_0 are respectively the apparent area of the specimen, the total ultimate force at fracture (maximum of the force-displacement curve), displacement at fracture and the initial height of the specimen.

2.4. FE Simulation

Twenty-three μFE analyses were performed on the twenty-three specimens to calculate compressive yield strengths. The μFE model constructed included a representation of rigid plates located on the upper and lower surfaces of the specimens (Fig. 2). The lower rigid plate was clamped and quasi-static compressive displacement was applied to the upper rigid plate with a displacement resolution of 0.01 mm till complete failure.

The force displacement curves were recorded for all the simulated tests and the ultimate apparent stress (σ_u) and strain (ε_u) were calculated using Eq. (6) and (7).

Contact between the upper and lower surfaces of the specimen and the plates was modeled using contact elements with a zero friction value to ensure that only compressive forces were transmitted.

Accurately estimating the tissue level properties of trabecular bone ultimate strain at fracture is difficult due to the complexity of the trabecular bone architecture (Niebur et al., 2000; Nagaraja et al., 2005). In all the analyses, the FE models were assigned the same tissue material properties, i.e. an isotropic elastic modulus of 18.0 GPa, and a Poisson's ratio of 0.3 (Sanyal et al., 2012).

(a) Quasi static compression load setup. (b) FE model reproducing the compression tests.

Figure 2. Boundary conditions applied to the specimens. (a) Experimental set up used to validate the proposed FE model for seven trabecular samples. (b) Same BC were applied to the other 23 remaining samples to perform the fracture simulations: Quasi-static compressive displacement was applied on the upper rigid moving plate (resolution of 0.01 mm). The lower rigid plate was clamped.

Very little information about yield and post-yield behavior of human trabecular tissue is available. Hence, failure models and their parameters are usually based on experimental data obtained for cortical bone (Hambli, 2012). In the current work, the ultimate strains in tension and compression of human cortical tissue were used (Parsamian, 2002).: $\varepsilon^{T}_{f-true} = 1.57\%$ and

$\varepsilon^{C}_{f-true} = 2.5\%$. The yield strain (ε_y) was set to ($\varepsilon_y = 0.7\%$) and the damage exponent to (*n*=2) (Parsamian, 2002; Hambli, 2012).

2.5. Model Parameters Sensitivity Analysis

Because of uncertainty concerning the fracture model parameters, a sensitivity analysis (SA) was performed to investigate the impact of the factors' sensitivities on the predicted ultimate stress and strain of the trabecular specimens. There are several different approaches to performing SA including one-factor SA and multi-factor SA. One-factor sensitivity analysis is useful in investigating the impact of one parameter varying in the model. In contrast, multi-factor SA examines the relationship between two or more different parameters that change simultaneously and their impact on the model responses. Multi-factor SA generally requires a huge amount of calculation time. In addition, it should be noted that the presentation and interpretation of multi-factor SA becomes increasingly complex as the number of parameters involved increases. The proposed trabecular fracture model is based on 7 model parameters (Young's modulus, Poisson ratio, strain at fracture in tension/compression and the critical damage value at fracture in tension and compression/shearing). Therefore, a limited preliminary one-factor SA analysis was performed in which only one model parameter value was varied by a given amount while the other parameters were kept at their reference values. For each parameter change, a fracture simulation was performed and the ultimate stress and strain was computed for each specimen. The fracture model input parameters listed in Table 2 were each varied by -15% and +15% with respect to their reference values (Cox et al., 2011; Hambli et al., 2012b).

A single run of the reference parameter values of the model was compared to single runs of the model with each parameter changed individually. SA analysis consisted of 6 runs for each specimen, totaling 138 runs. The total computation time was about 15 hours on a 64 GB computer.

3. Results and Discussion

The current FE trabecular bone fracture model provided excellent agreement between predicted and experimentally measured values of apparent ultimate stress (R^2=0.98) and apparent ultimate strain (R^2=0.93) in compression performed on 7 specimens (Figure 3).

An example of a predicted trabecular specimen fracture pattern is compared to the experimental one consisting in a 3D microCT scan of the specimen before and after compressive mechanical test till complete failure (Figure 4).

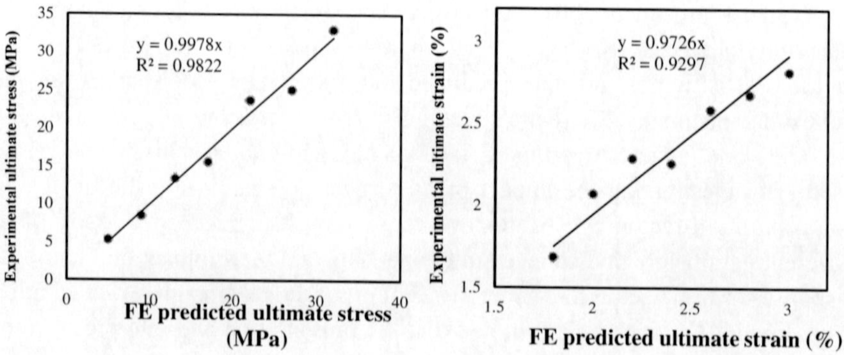

Figure 3. Validation of predictions of compressive ultimate stress and strain obtained for 7 specimens.

(a) Rendered experimental initial state. (b) Rendered experimental failed state. (c) Prediction failed state.

Figure 4. Example of a 3D reconstruction obtained by microCT-scan of a trabecular bone specimen before (a) and after the compressive mechanical test at fracture (b). The FE predicted broken specimen is rendered with colors representing the completely damaged trabeculae (red) and Non damaged tissue (grey) (c). Predicted fracture pattern agrees well with the experimental one concerning the localization of the fracture (failure occurring in oblique manner as indicated by the blue arrows).

Applied compressive load resulted in localized fracture features (damage/cracks) in the specimen following an oblique cracking path. Very similar fracture pattern was obtained by the FE model indicating the current fracture approach is able to predict both force-displacement curves and fracture patterns of cancellous bone specimens.

A typical experimental force-displacement curve of the tested specimens is given in Fig. 5. The curves show the same trend for curve shape and onset of yielding and fracture as the experimental one. The FE-based curve exhibits a drop in force during failure that was nearly always observed during mechanical testing. After having reached the ultimate force at fracture, the force decreases gradually for increasing displacement, which corresponds to a softening stage till complete fracture.

Deformation starts with an elastic stage (*A*). In the yielding stage (*B*), the crack is initiated at a limited number of trabeculae. After the yielding phase, the crack continues to grow rapidly to reached the ultimate stress at fracture (*C*) leading to a softening stage till complete fracture. This phase corresponds to the generation of microcracks till complete fracture of the specimen (*D*). The damage threshold stress which coincides with the elastic limit or yield stress corresponding to the yield strain (ε_y) (*B*) quantifies the onset of damage behavior.

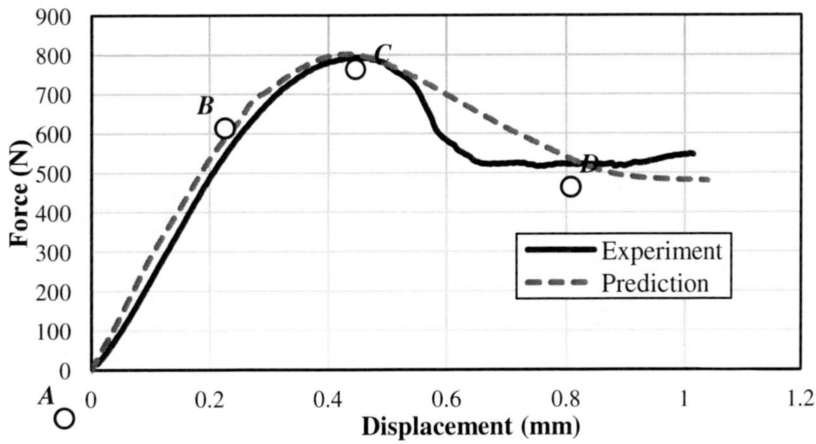

Figure 5. Force displacement curve. Comparison between experiment and FE prediction.

The results indicate that local damage in trabecular bone initiates at 0.3% apparent strain. This finding implies that tissue yielding can initiate at very low apparent strains in the trabecular bone. This local failure has negative consequences on the apparent fracture properties of trabecular bone.

At large apparent strain, 25%–40%, the progressive contact of trabeculae generate the compaction of the specimen microstructure which lead to increase of the load carrying capacity.

This phenomenon occurs after the specimen fracture and therefore, the self contact between the trabeculae is not modeled in the current work for simplicity. An example of the propagation of damage within a specimen in relation to the force-displacement curve position is plotted in Figure 6.

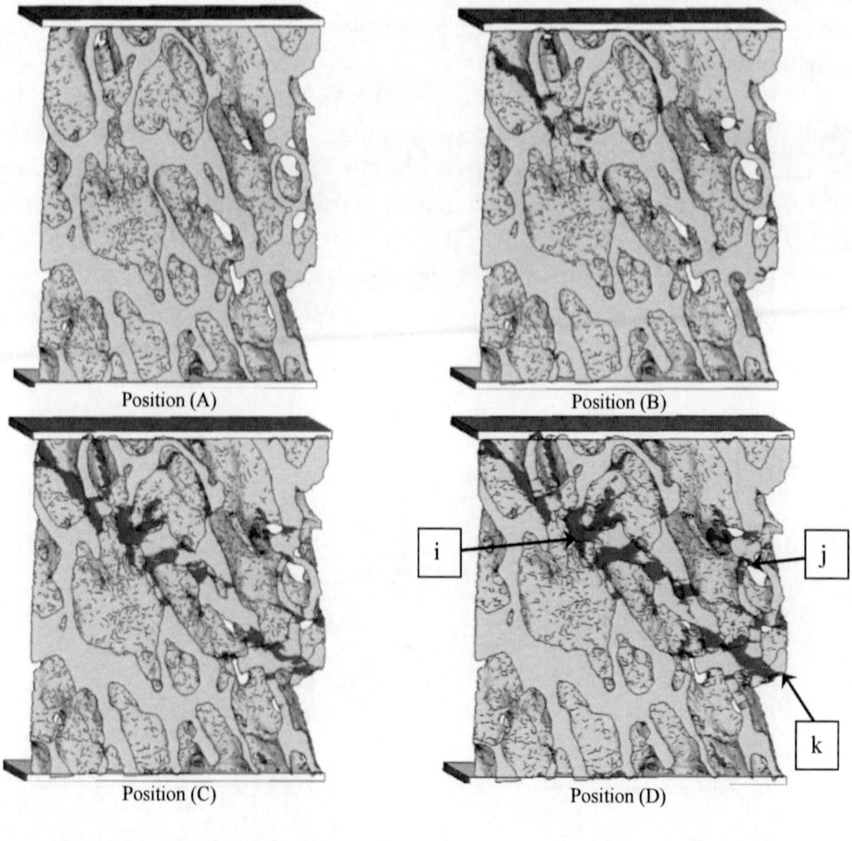

Figure 6. Damage propagation sequences (A to D) in a longitudinal cross section in relation to the force-displacement curve and different modes of local bone tissue failure under compression: Bending (i), Buckling (j) and Shearing (k).

Finite element analysis revealed that damage was concentrated in microfractured trabeculae. Failures in the longitudinal trabeculae were predominantly generated by compressive deformation, leading to buckling failure, whereas failure in the oblique/horizontal trabeculae was predominant and was generated by bending and shear.

The results shown in Figure 6 indicate that the predicted fracture process occurred in three steps: (i) development of initial cracks in the trabeculae regions subjected to a local maximum tensile stress; (ii) propagation of the cracks within these trabeculae in the direction of the local maximum compressive stress, and (iii) the complete failure mechanism of the specimen is the result of the failure propagation path of a small percentage of trabeculae corresponding to the regions dominated by the shearing stress.

The damage contours for the two meshes (coarse and fine) and corresponding force-displacement curves are given in Figure 7 (A) and (B).

Predicted damage distribution using the proposed mesh regularization technique reveals that the contours are very similar indicating that the mesh dependency is not significant. It can be observed that the fine mesh generated a localized damaged region on the trabeculae compared to coarse one.

Little difference can be observed concerning the force-displacement curves during the post fracture behavior. Nevertheless, for fine mesh, a rapid drop of the curve after the occurrence of the first cracks indicating that fine meshes could lead to an increase of the softening rate during the fracturing process (cracks propagation speed increases with the mesh refinement). This results reveals a certain mesh dependence of the solution remain. Nevertheless, this limited effect have no influence on the predicted apparent ultimate stress and strain at fracture (maximum of the curve and onset of the fracture are not changed).

Taken together, the mesh dependency analysis showed that the overall mechanical damage and fracture behavior of the trabecular bone samples was not affected by mesh size suggesting that the proposed mesh regularization technique is reliable.

The predicted apparent compressive ultimate stress for the different trabecular specimens versus varying bone volume fraction (BV/TV) is plotted in Figure 8.

Similar results were obtained between the current prediction and previously published experimental results (fitted law provided by Hernandez et al., 2001) and FE results reported recently by Sanyal et al. (2012).

(a) Coarse mesh (525494 FE elements). (b) Fine mesh (1202120 FE elements).

(A) Damage contour.

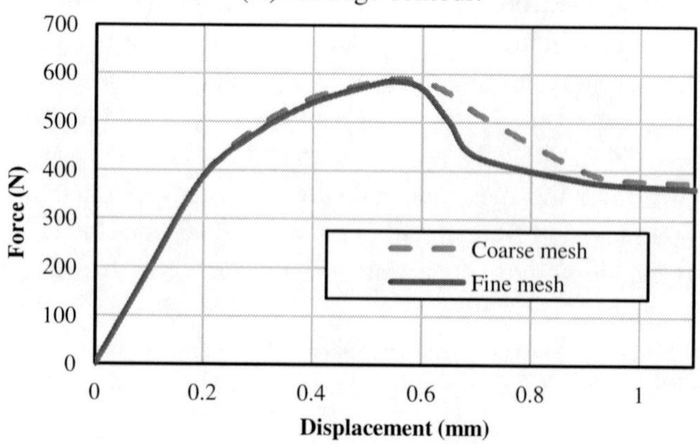

(B) force–displacement curves.

Figure 7. (A) Initial meshes and predicted damage contours corresponding to the complete fracture of the specimen obtained by coarse (a) and fine (b) meshes. (B) Predicted force–displacement curves.

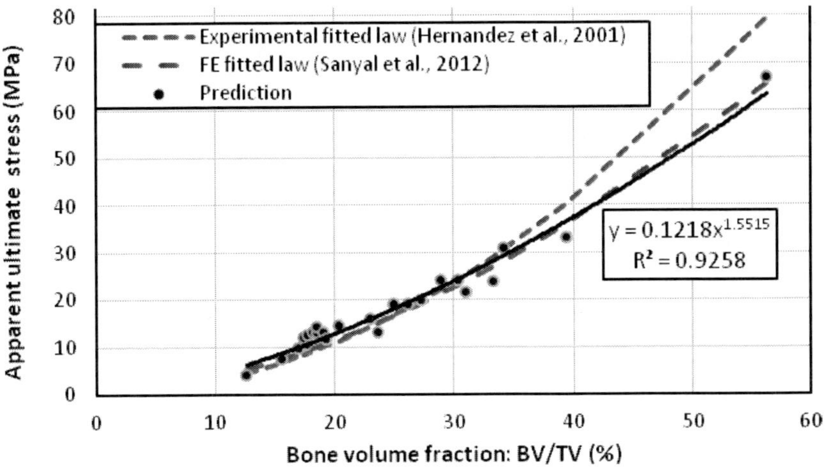

Figure 8. Variation in apparent compressive ultimate stress versus bone volume fraction (BV/TV). Current prediction (dots), experimental fitted law from Hernandez et al. (2001) (blue dashed line), FE fitted law from Sanyal et al. (2012) (red dashed line) and current predicted fitted law (solid black line).

The predicted results suggest that the decrease in the ultimate stress of trabecular bone is mainly a consequence of the loss of trabeculae, rather than a variation in the material properties of the trabecular bone tissue. In contrast, the limited dependence of the ultimate strain on the bone volume fraction may indicate that ductility/fragility of bone is related to other factors such as the collagen concentration in the bone which has been reported to be an accurate predictor of ultimate strain (Ding et al., 1997; Ding and Hvid, 2000). Biomechanical studies have also shown a strong correlation between the trabecular bone density and the whole bone strength (Lotz and Hayes, 1990).

Figure 9 plots the variation of the predicted apparent ultimate strain versus the bone volume fraction.

It can clearly observed that the ultimate strain at fracture exhibits a slight non-linear increase with increase in bone volume fraction. This indicates that age, gender or site-related differences in apparent trabecular morphology may have only a limited effect on apparent yield strains. This is in agreement with the results reported by (Kopperdahl and Keaveny, 1998). Similar results were reported recently by Sanyal et al. (2012). The authors applied an experimentally validated non-linear FE model to investigate the fracture behavior of trabecular bone specimens. They found that the apparent-level yield strain increases with bone volume fraction for compressive loading.

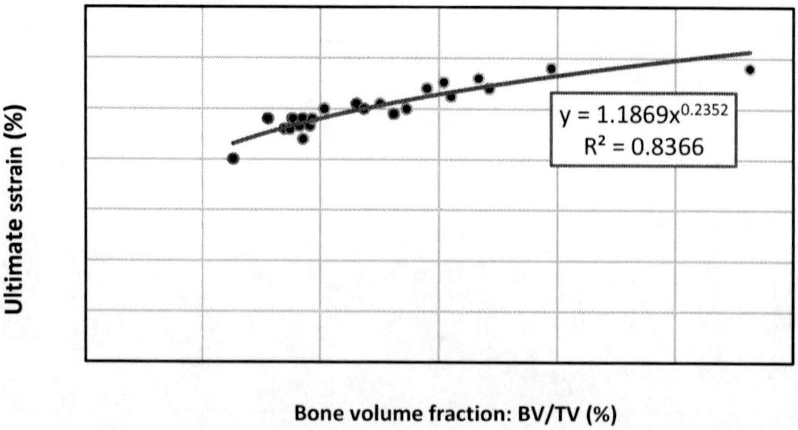

Figure 9. Variation in predicted apparent compressive ultimate strain versus bone volume fraction (BV/TV).

The current predicted results indicate that the coefficient of variation of the apparent ultimate strain is small (10%) and that the apparent ultimate strain exhibits two regimes related to the bone volume fraction ranges. Regime (i) corresponds to (BV/TV<35%) where the ultimate strain increases slightly (gentle slope) and Regime (ii) corresponds to (BV/TV>35%) where the variation in the apparent-level ultimate strain values is negligible and can be considered uniform: 2.8 ± 0.1.

Contrary to the ultimate stress, Fig. 8 indicates that trabecular bone ultimate strain is relatively homogeneous despite the variation in bone volume fraction, age and gender of the specimens. This result agrees well with experiments performed by several authors showing that the ultimate strain of trabecular bone has only a weak dependence on apparent density (Rohl et al., 1991; Keaveny et al., 1994; Kopperdahl, and Keaveny 1998 ; Hou et al., 1998). These findings imply that applying a failure criterion in the FE models to simulate bone organ fracture based on isotropic inhomogeneous ultimate strain depending on the trabecular bone site can generate accurate results. In this case, FE modeling of the trabecular region of a whole bone organ can be performed using continuum FE models with spatially variable material properties depending of the architecture on the trabecular bone (material properties assigned to each FE based on the local micro-architecture properties such as bone volume fraction). Such continuum FE models are computationally far less demanding. Nevertheless, multiscale modeling approaches are still needed to reflect the real microstructure.

Note that the mean apparent ultimate strain at fracture (2.5 ± 0.5) is greater than the strain at fracture (ε_f) of the average bone tissue (2%). This indicates that compressive load on trabecular specimens induced a coordinated deformation process at the trabeculae level. These findings suggest that the architecture of trabecular bone leads to a gradation in the compressive strain applied to the trabeculae of the whole specimen. Trabecular deformation generates a damping process mainly by bending of the oblique trabeculae, which increases the ductility of the whole specimen (its ability to sustain deformation).

Variation in the ultimate stress and strain versus age is given in Figure 10. The predicted results show that the ultimate stress declines (R^2=0.5776) with ageing and that no significant gender difference is observed. In contrast, no correlation between the ultimate strain and age was found (R^2=0.0184). Very similar experimental results were reported by (Lindahl, 1976.; Mosekilde et al., 1987; McCalden et al., 1997; Ding et al., 1997). These previous studies have reported (i) a significant reduction in the ultimate stress of trabecular bone versus age and (ii) no variation in the ultimate strain of trabecular bone with age. Ding et al. (1997) tested in compression specimens of human proximal tibial trabecular bone from 31 normal donors aged from 16 to 83 years and determined the ultimate stress and strain at fracture. The authors suggested that the decrease in the ultimate stress is mainly a consequence of the loss of bone volume fraction and collagen concentration. Current predicted results agree very well with these experimental findings, indicating that the current FE model may be used to investigate the role of physiological and mechanical trabecular bone parameters on its apparent fracture properties.

No significant differences between men and women were predicted, in agreement with previous studies suggesting that after about 35 years old the fracture properties of bone deteriorate in both men and women (Burstein et al., 1976; Smith and Smith, 1976; Currey, 1979; McCalden et al., 1993). During ageing, the net loss of trabecular bone is less in men than in women (Duan et al., 1999, 2001; Seeman, 2002). For women, menopause-related oestrogen deficiency accelerates bone loss and causes trabecular thinning and disconnection. Bone thinning predominates in men rather than disconnection (Seeman, 2002). As a result, women are more prone than men to osteoporosis which generates a more pronounced decline of the trabecular bone volume fraction, and explains the predominance of bone organ fracture for women compared to men (Aaron et al., 1987). The independence of the predicted apparent ultimate strain of age and gender for the current specimens should be studied by considering other trabecular bone factors and lower-scale tissue level properties to investigate the nature of gender differences in apparent bone fragility/ductility.

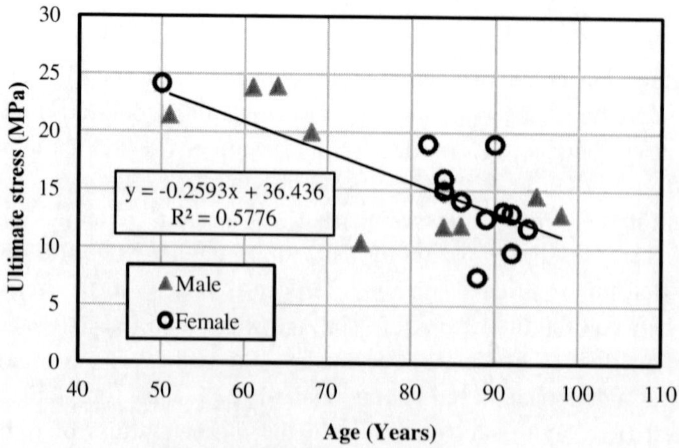

(a) Ultimate stress (MPa).

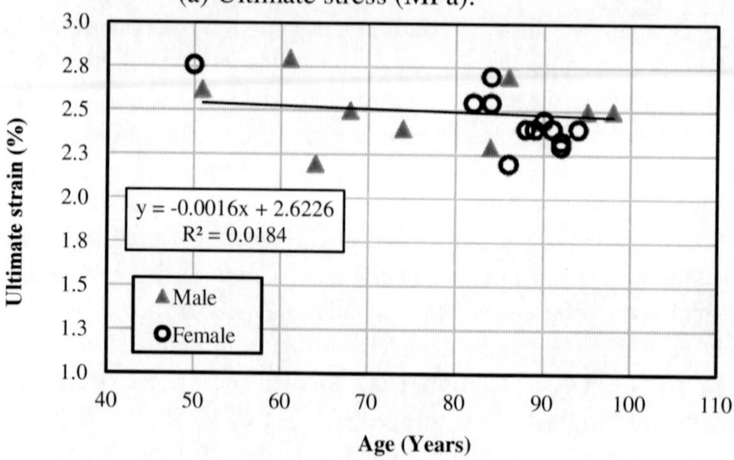

(b) Ultimate strain (%).

(black circles: female (n=12), red triangles: male (n=9)).

Figure 10. The ultimate stress and strain of human femoral trabecular bone versus age.

The energy absorbed at failure is the area under the stress-strain curve and is known as the toughness. Toughness of bone is its ability to resist fracture by absorbing mechanical energy. The specimens' toughness was predicted by calculating the areas under the apparent stress-strain curves until failure point.

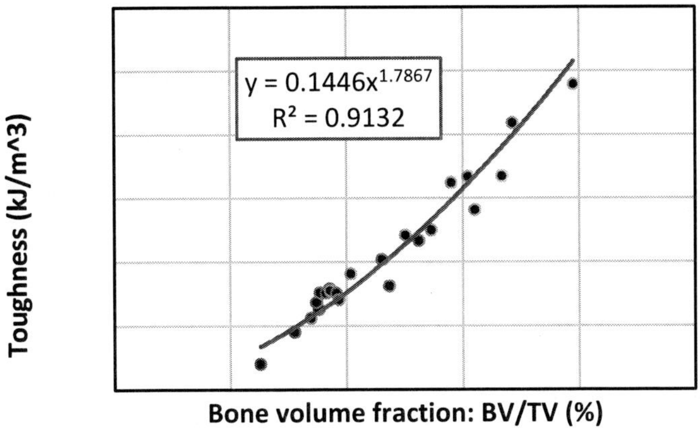

Figure 11. Trabecular bone toughness versus BV/TV predicted by the current FE model.

The results reveal that toughness is highly dependent on bone volume fraction (Fig. 11) due to the net increase in the ultimate stress versus BV/TV. Toughness increased by a power law relationship with volume fraction (BV/TV) (R^2= 0.91). Our predicted results are consistent with those of previously published studies (Yeni et al., 1997; Zioupos, 1998; Garrison et al., 2009).

The determinant of apparent bone strength is the properties of the elementary constituent material (Barkaoui et al., 2011; Hambli and Barkaoui, 2012). At the material level, trabecular bone tissue is a composite of an organic matrix containing primarily Type I collagen and an inorganic phase of crystalline mineral. The characteristics of the two phases and their interactions/organization determine the material properties of bone tissue. At macroscale, bone mineral content is directly related to the bone stiffness and strength (Burstein et al., 1976; Currey, 1988; Battaglia et al., 2003). At the micro- and nanoscale, mineral characteristics such as crystallinity and carbonate substitution determine material properties that contribute directly to whole bone stiffness and strength, and collagen characteristics in the extracellular matrix, such as collagen content and maturity or crosslinking, contribute primarily to the toughness and postyield deformation of the tissue (Burstein et al., 1976; Currey, 1988; Boskey et al., 1999, Burr, 2002). As trabecular bone architecture is a natural cellular solid, the deformation mechanisms of the trabecular specimens depend on the open/closed architecture (disconnected/connected trabeculae) of the cell material. The consequence of such a solid-like cell is that the apparent strength properties of the structure (ultimate stress and toughness) follow a power function with respect to bone volume fraction material from 1.5 to 2 (Figs 8 and 11).

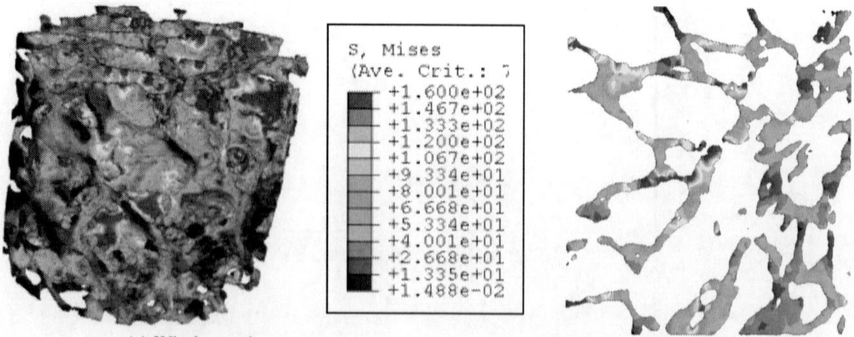

(a) Whole specimen. (b) A longitudinal cross section.

Figure 12. Contour of von Mises stress and stress concentration in some trabeculae. (a) Whole specimen. (b) A longitudinal cross section. Apparent applied stress=40 MPa, Maximum localized stress=155 MPa.

Rapid growth of damage can be explained by the existing local non-linearities generated by the stress concentration (Figure 12). This stress concentration can generate failure due to plastic damage accumulation in localized regions prior to apparent yield stress and the apparent mechanical properties degenerate in the localized regions. Our results are consistent with previous studies (Nagaraja et al., 2005, Arthur Moore and Gibson, 2002).

At the tissue level, it was found that the trabeculae are locally subjected to mixed strain modes including tension, compression and shearing depending on the oblique orientation of the trabeculae (Figure 13).

Figure 14 shows the trabeculae failure (in red) corresponding to complete fracture of four different specimens with different BV/TV values.

It was found that damaged trabeculae in the specimens possessed significantly higher local stresses and strains than undamaged trabeculae and that complete damage is localized to these limited number of trabeculae.

In order to account for the fracture in the bone samples, an example of damage distribution in 8 different subdivided cross sections segments corresponding to the complete specimen fracture is given in Figure 15. As explained earlier, $0<D<1$, with $D=0$ indicating non damaged bone tissue and $D=1$ signifying complete local damaged or fractured bone site. D is therefore a state variable quantifying the degree of the damage, providing a phenomenological measure for failure analyses.

Figure 13. Distribution of tissue level compressive, tensile and shear strains (BV/TV=33.67%). Red (compression), Blue (tension) and green (shearing).

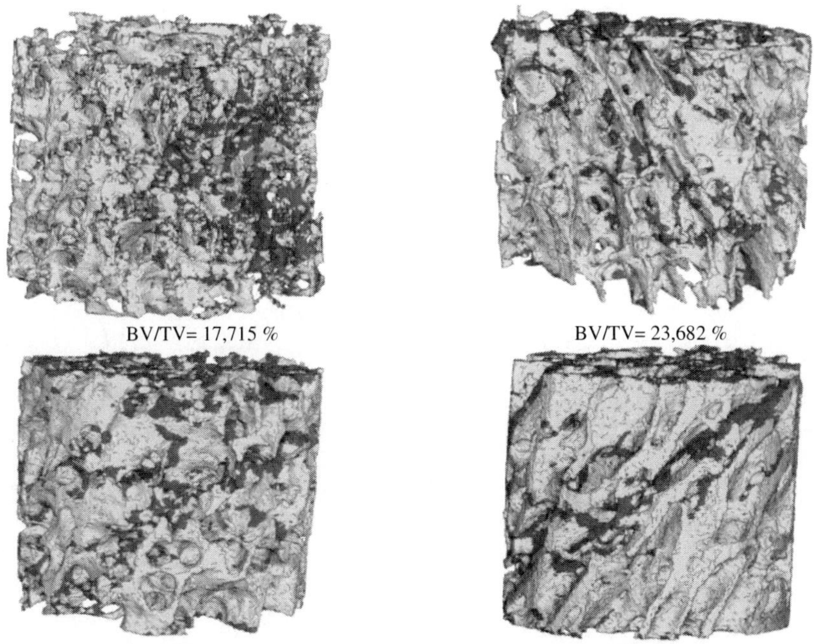

Figure 14. Distribution of tissue-level fracture (cross section) for four different specimens with low (BV/TV=12.612 %), medium (BV/TV=19.330 %) and high (BV/TV= 39.482 %) values. (Red: completely damaged trabeculae).

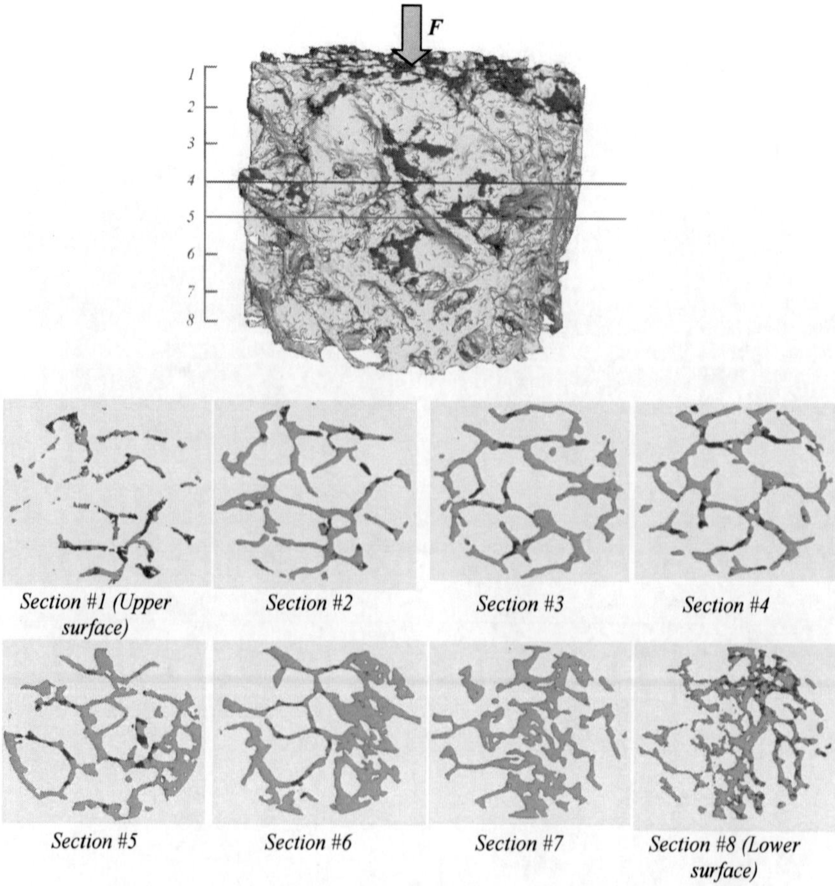

Figure 15. Damage distribution in 8 different subdivided cross sections (equal-sized) at the end of the loading. (Red: Completely damaged bone tissue), (Grey: Non damaged tissue). The results indicate that the trabecular architecture induced an inhomogeneous distribution of damage in the trabeculae (bone tissue) and that the damage increases in the region of low BV/TV.

The variation of the number of the broken trabeculae versus BV/TV at every bone segments is plotted in the form of histogram (Fig. 16). The results indicate that the trabecular architecture induced an inhomogeneous distribution of damage in the trabeculae (bone tissue) and that the damage is localized in the region of low BV/TV. This result is in agreement with the findings of Nazarian et al. (2008) who reported that bone tissue deformation was the highest in the bone segments with the lowest BV/TV. This result is in conformity with the finding of Perilli, et al. (2006) and Nazarian et al., (2008).

The authors reported that the bone tissue deformation was the highest in the bone segments with the lowest BV/TV.

The results indicate that bone slices with low bone volume (a lower number of trabeculae) are subjected to a higher damage concentration effect. Even if the apparent behavior of the specimen appears to be linear, non-linearities can be found locally due to strain concentration. This is in agreement with the experiments by Dendorfer et al. (2008). The authors showed that accumulated residual and maximum strains within trabecular bone specimens generate strain concentrations followed by the formation of a fracture line through the specimen. Martin et al. (1998) reported that relatively small amounts of microdamage could cause significant reductions in the bone's mechanical properties even before the appearance of microcracks. Similar results were obtained with numerical studies, which demonstrated that damage is visible below apparent compressive yield strain and that local material yielding initiates at low apparent stress levels (Nagaraja et al., 2005; Morgan et al., 2005).

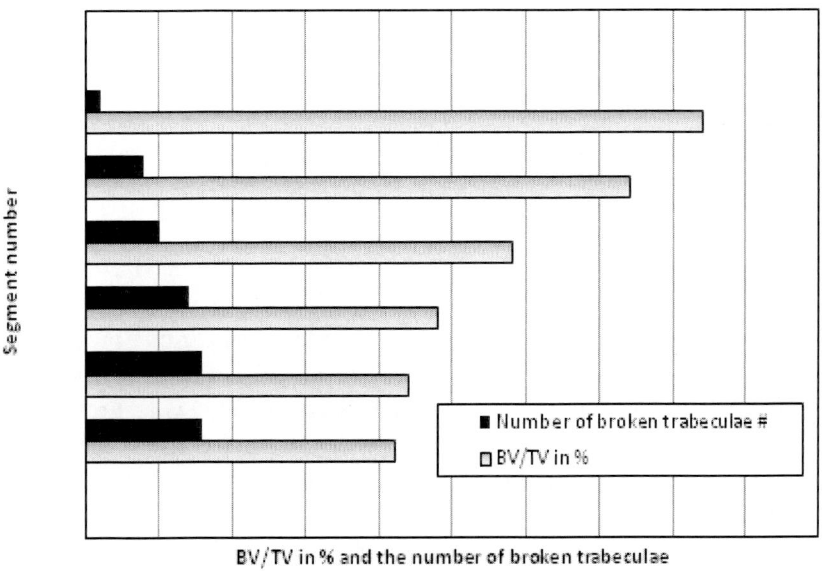

Figure 16. Variation of number of broken trabeculae versus BV/TV for the different trabecular bone segments (2 to 7). The segments 1 and 8 were omitted to avoid results artifacts due to excessive compression of the FE meshes contacting the upper and lower rigid plates.

3.1. Sensitivity Analysis

The averaged results of the SA obtained from the 23 specimens are plotted in Figure 17 which shows the impact that a fixed change in each model parameter has on the ultimate stress and strain.

The results of the SA underscore the influence of each parameter on the resulting ultimate stress and strain. Variation in the model parameters related to bone fracture properties revealed their significant influence on the ultimate stress/strain. This can be related to the fact that these properties explicitly affect the bone tissue resistance to fracture (onset and propagation).

SA analysis showed that (i) variation in the Young's modulus has a limited influence compared to the fracture properties and that (ii) the Poisson ratio has a negligible impact.

The impact of the mesh dependency mesh regularization model factors (L_{FE}) and (L_{frx}) were less than 0.2% . This results reveals a very limited impact if the mesh dependence factors sensitivity suggesting that the proposed mesh regularization technique is sufficient reliable.

The predicted results revealed that in all SA model cases, the fracture patterns were very similar, and were not significantly affected by variation in the parameters. Nevertheless, the rate of crack propagation increased with increasing bone fragility (decrease of bone tissue strain at fracture and critical damage at fracture), suggesting that bone fracture may range from brittle to quasi-brittle patterns depending on its tissue material properties.

Decreasing the strain at fracture and critical damage at fracture generated bone fragility and hence, significantly decreased the apparent fracture properties, while increasing those parameters increased the fracture force. From a material point of view, it is also important to note that the material properties needed for the current simulations can exhibit statistical variation (uncertainty). It is well established that the degree of mineralization, fatigue microcracks and material properties strongly influence the mechanical stiffness and strength of bone which can be related to ultimate stress (Morgan et al., 2001).

In summary, variation in the model parameters related to bone tissue fracture properties has a significant influence on the predicted apparent ultimate stress and strain force of trabecular bone and negligible affects on the fracture patterns for the current specimens. For future general SA analysis, it is necessary to consider the full factorial parameters variation simultaneously for different specimen geometries.

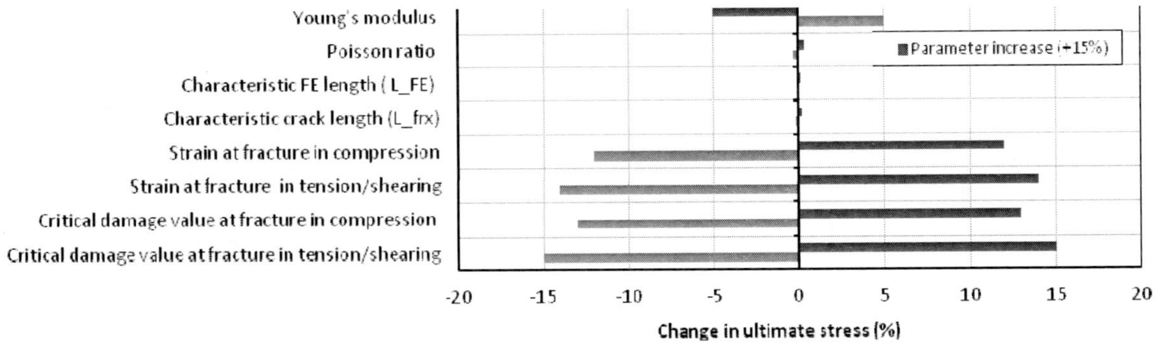

(a) Impact of a fixed change in each FE model parameter on the ultimate stress.

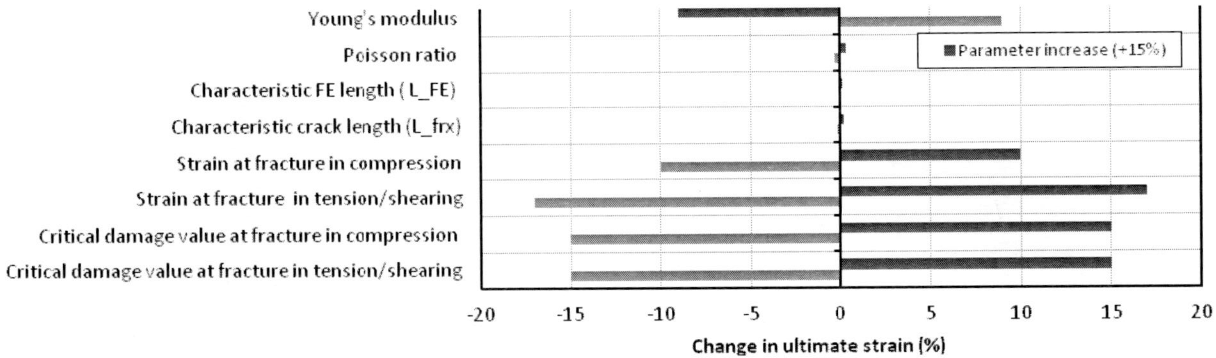

(b) Impact of a fixed change in each FE model parameter on the ultimate strain.

Figure 17. Relative percentage change of the ultimate stress and strain predicted by the trabecular specimens fracture FE simulations. Model parameters were varied between -15% and +15%. Sensitivity analysis was performed on the 23 specimens and averaged results are plotted.

Most micro-CT finite element analyses of this type of bone focused on linear and non-linear analysis of the bone's mechanical behavior in order to identify some mechanical properties and investigate the interactions between the morphological structures of the bone.

In the current study, an isotropic μFE model coupled to a damage law was developed in order to simulate fracture of human trabecular bone specimens under quasi-static compressive load and predict the apparent stress and strain at fracture and the final fracture pattern of the specimens. The element deletion technique was applied in order to simulate the progressive fracturing process of bone. The model was first validated with experimental results performed on 7 specimens, and was then applied to assess the apparent stress and strain at fracture of 23 human trabecular bone specimens. The simulation results were compared with reported experimental results to investigate the accuracy of the μFE model presented here. It was found that: (i) the behavior of trabecular bone subjected to compressive mechanical load is characterized by initiation and growth of damage that later localizes in relatively narrow fracture process zones. (ii) Predicted failure properties agree very well with the experimental ones. (iii) BV/TV is a very good predictor of the bone fracture properties under quasi-static loads. In addition, it was found that the trabecular architecture induces an inhomogeneous distribution of damage in the trabecula. (iv) The predominant compression failure mode appears to be a combination of bending, buckling and shearing of individual trabeculae. (v) Taken together, the predicted results of the current study show that variations in the apparent fracture properties (ultimate stress, strain and toughness) of trabecular bone are strongly related to variation in the bone volume fraction. However, despite the success of the proposed trabecular bone fracture FE modeling approach, certain limitations apply. First, for simplicity's sake, the FE model of the trabecular bone was considered isotropic and material data were obtained from published experimental results on human cortical trabecular bone tissue due to lack of knowledge related to trabecular tissue. Since the investigation scale of the present work corresponds to one or some trabeculae levels, one can assume isotropic averaged properties. However, limited experimental data are available to characterize the micromechanical properties of trabecular bone.

In addition, it has been reported that bone exhibits complex behavior including anisotropy, viscosity, and inelastic deformation. Recently, several anisotropic viscoelastic behavior laws coupled to damage for bone tissue level have been developed by various authors (Natali et al., 2008; Garcia et al., 2009). However, the main limitations in the application of these sophisticated models is that they need a large number of parameters (anisotropy, viscosity and damage) that are not available in the published literature and are very complicated to measure experimentally. Second, although the present FE model was well validated for compressive ultimate stress performed on 7 specimens, extensive validation is necessary in tension, shearing and mixed deformation modes. Third, the harvested specimens do not account for the variation of other factors reported to be important, such as mineral density, tissue material properties and the geometric detail of the trabeculae. Therefore, the conclusions of the current study are limited to the case of the current specimens. More variants of trabecular specimens are needed to capture the effects of biomechanical factors which affect apparent bone behavior. Nevertheless, the overall structure of the proposed FE approach will remain unchanged. There will still be a need to perform FE simulations to investigate the relationship between the apparent fracture properties of human trabecular bone and bone biomechanical characteristics. A potential source of uncertainty in the FE modeling was the application of the same material properties to each trabecular specimen harvested from different bone sites. Also, there are several aspects of damage growth in trabecular bone prior to complete failure that remain poorly understood. The relationship between changes in mechanical properties (residual strain, anisotropy) and damage accumulation (type and amount) is not yet well defined. Despite these limitations, the proposed micro-CT FE model should provide an improved approach for predicting the ultimate stress and ultimate strain of trabecular bone which is needed to perform FE simulations in biomechanical analyses of whole bone organs such as femurs.

Acknowledgments

This work was supported by the French National Research Agency (ANR) through the TecSan program (Project MoDos, n°ANR-09-TECS-018). The author is grateful to Dr Gassara F. for providing the micro-CT trabecular bone meshes and experimental results.

References

Aaron J.E., Makins N.B., Sagreiya K., 1987, The microanatomy of trabecular bone loss in normal aging men and women. *Clin Orth RR*; 215: 260–71.

Abaqus (6.11), 2012, *Documentation, Analysis User's Manual.*

Abdel-Wahab A.A. and Silberschmidt V V. 2011, Numerical modeling of impact fracture of cortical bone tissue using X-FEM, *J. of theoretical and applied mechanics*, 49:3, 599-619.

Arthur Moore, T.L., Gibson, L.J., 2002. Microdamage accumulation in bovine trabecular bone in uniaxial compression. *Journal of Biomechanical Engineering* 124 (1), 63–71.

Barkaoui A., Bettamer A. and Hambli R., 2011, Failure of mineralized collagen microfibrils using finite element simulation coupled to mechanical quasi-brittle damage, *Procedia Engineering* 10: 3185-3190.

Battaglia T.C., Tsou A.C., Taylor E.A., Mikić B., 2003, Ash content modulation of torsionally derived effective material properties in cortical mouse bone. *J Biomech Eng.*;125:615–619.

Bayraktar H.H. and, Keaveny T.M., 2004, A computational investigation of the nonlinear behavior of human trabecular bone., *Transactions of the 12th annual Pre-ORS Symposium on Computational Methods in Orthopaedic Biomechnics*, 2.

Bayraktar H.H., Morgan E.F., Niebur G.L., Morris G.E, Wong E.K. and Keaveny T.M., 2004, Comparison of the Elastic and Yield Properties of Human Femoral Trabecular and Cortical Bone Tissue, *Journal of Biomechanics,* 37(1):27–35.

Bevill, G., Eswaran, S.K., Gupta, A., Papadopoulos, P., Keaveny, T.M., 2006. Influence of bone volume fraction and architecture on computed large-deformation failure mechanisms in human trabecular bone. *Bone* 39 (6), 1218–1225.

Bevill, G., Farhamand, F., Keaveny, T.M., 2009. Heterogeneity of yield strain in low-density versus high-density human trabecular bone. *Journal of Biomechanics* 42 (13), 2165–2170.

Boskey A.L., Wright T.M., Blank R.D., 1999, Collagen and bone strength. *J Bone Miner Res*, 14:330–335.

Budyn E., Hoc T. and Jonvaux J., 2008, Fracture strength assessment and aging signs detection in human cortical bone using an X-FEM multiple scale approach, *Comput Mech* 42:579–591.

Burr D.B. 2002. The contribution of the organic matrix to bone's material properties, *Bone* 31:8–11.

Burstein, A.H., Reilly, D.T. and Martens, M.,1976, Aging of bone tissue: mechanical properties. *Journal of Bone Joint Surgery* 58-A, 82–86.

Cox L.G., van Rietbergen B., van Donkelaar C.C., Ito K., 2011, Analysis of bone architecture sensitivity for changes in mechanical loading, cellular activity, mechanotransduction, and tissue properties. *Biomech Model Mechanobiol.* 10(5):701-12.

Currey, J.D.,1979, Changes in the impact energy absorption of bone with age. *Journal of Biomechanics* 12, 459–469.

Currey J.D., 1988, The effect of porosity and mineral content on the Young's modulus of elasticity of compact bone. *J Biomech.*;21:131–139.

Currey J.D., 1990, Physical characteristics affecting the tensile failure properties of compact bone. *Journal of Biomechanics*, 23(8):837–844.

Currey J.D., 2002. *Bones: Structure and Mechanics*. Princeton University Press, Princeton.

Dendorfer S., Maier H.J. and Hammer J., 2008, Anisotropy of the fatigue behaviour of cancellous bone, *J. Biomech* 41 (3), 636–641.

de Vree, J.H.P., Brekelmans, W.A.M. and van Gils, M.A.J., 1995. Comparison of nonlocal approaches in continuum damage mechanics. *Comp. Struct.*, 55, 581–588.

Ding M., Dalstra M., Danielsen C.C., Kabel J., Hvid I., Linde F., 1997, Age variations in the properties of human tibial trabecular bone. *J Bone Joint Surg Br.*;79(6):995-1002.

Ding M, Hvid I., 2000, Quantification of age-related changes in the structure model type and trabecular thickness of human tibial cancellous bone. *Bone*;26:291–5.

Duan Y., Parfitt M., Seeman E., 1999, Vertebral bone mass, size and volumetric bone mineral density in premenopausal women, and postmenopausal women with and without spine fractures. *J Bone Miner Res*; 14: 1796–1802.

Duan Y., Turner C.H., Kim B.T., Seeman E., 2001, Sexual dimorphism in vertebral fragility is more the result of gender differences in age-related bone gain than bone loss. *J Bone Miner Res*; 16: 2267–75.

Fondrk, M.T., Bahniuk, E.H., Davy, D.T., Michaels, C. 1988. Some viscoplastic characteristics of bovine and human cortical bone. *J. Biomech.* 21(8), 623–630.

Fondrk, M.T., Bahniuk, E.H., and Davy, D.T. 1999. Inelastic strain accumulation in cortical bone during rapid transient tensile loading. *Journal of Biomechanical Engineering*, 121:616–621.

Garcia D, Zysset P, Charlebois M, Curnier A. 2009. A three-dimensional elastic plastic damage constitutive law for bone tissue. *Biomech Model Mechanobiol.* 8(2):149–165.

Garrison J.G., Slaboch C.L., Niebur G.L. , 2009, Density and architecture have greater effects on the toughness of trabecular bone than damage. *Bone*;44:924–9.

Hambli R., 2012, A quasi-brittle continuum damage finite element model of the human proximal femur based on element deletion, *Med. & Biol. Eng. & Comp.* 51:1-2, 219-231.

Hambli R, Bettamer A, Allaoui S. 2012a, Finite element prediction of proximal femur fracture pattern based on orthotropic behaviour law coupled to quasi-brittle damage. *Med Eng Phys.* 34(2):202-10.

Hambli R., Lespessailles E. and Benhamou C.L., 2012b, Integrated remodeling-to-fracture finite element model of human proximal femur behaviour, *J Mech Behav Biomed Mater* (In press, available online).

Hambli R. and Barkaoui A., 2012, Physically based 3D finite element model of a single mineralized collagen microfibril, *Journal of Theoretical Biology* 30:128-141.

Harrison NM, McDonnell P, Mullins L, Wilson N, O'Mahoney D, McHugh PE (2012) Failure modelling of trabecular bone using a non-linear combined damage and fracture voxel finite element approach. *Biomech Model Mechanobiol* 1–17. doi:10.1007/ s10237-012-0394-7

Hernandez CJ, Beaupré GS, Keller TS, Carter DR., 2001, The influence of bone volume fraction and ash fraction on bone strength and modulus, *Bone* ;29(1):74-78.

Hosseini H.S., Pahr DH.., Zysset P.K., 2012, Modeling and experimental validation of trabecular bone damage, softening and densification under large compressive strains, *J Mech Behav Biomed Mater.* 15:93-102.

Hou, F.J., Lang, S.M., Hoshaw, S.J., Reimann, D.A., and Fyhrie, D.P., 1998, Human vertebral body apparent and hard tissue stiffness, *J. Biomech.*, 31, 1009–1015.

Huang Z., Ryu W., Ren P., Fasching R., Goodman S.B., 2008, Controlled release of growth factors on allograft bone in vitro. *Clinical orthopaedics and related research*, 466:1905-1911.

Jungmann R, Szabo ME, Schitter G, Tang RY, Vashishth D, Hansma PK, Thurner PJ., 2011, Local strain and damage mapping in single trabeculae during three-point bending tests, *J. Mech. Behav. Biomed. Mater.*, 4(4):523-34.

Kabel, J., van Rietbergen, B., Dalstra, M., Odgaard, A., and Huiskes, R., 1999, The Role of an Elective Isotropic Tissue Modulus in the Elastic Properties of Cancellous Bone, *J. Biomech.*, 32:673–680.

Kaneko T.S., Pejcic M.R., Tehranzadeh J., Keyak J.H., 2003, Relationships between material properties and CT scan data of cortical bone with and without metastatic lesions, *Med Eng Phys*., 25(6):445-54.

Keaveny T.M., Wachtel E.F., Ford C.M. and Hayes W.C., 1994. Differences between the tensile and compressive strengths of bovine tibial trabecular bone depend on modulus. *Journal of Biomechanics* 27,1137-1146.

Keaveny T.M. Wachtel E.F. and Kopperdahl D.L., 1999, Mechanical behavior of human trabecular bone after overloading. *Journal of Orthopaedic Research,* 17:346–353.

Keaveny T.M., Morgan E.F., Niebur G.L. and Yeh, O.C., 2001, Biomechanics of trabecular bone. *Annual Review of Biomedical Engineering*, 3:307–333.

Kopperdahl D.L. and Keaveny T.M., 1998, Yield strain behavior of trabecular bone, *Journal of Biomechanics* 31, 601-608.

Kotha S.P. and Guzelsu N., 2003, Tensile damage and its effects on cortical bone. *Journal of Biomechanics*, 36(11):1683–1689.

Kinney J.H. and Stölken J.S., 2005, The importance of geometric nonlinearity in finite element studies of yielding in trabecular bone. *Trans Orthop Res Soc* 28:28.

Lemaitre J., 1985, A continuous damage mechanics model for ductile fracture. *Journal of Engineering Materials and Technology*. 107:83-89.

Lindahl O., 1976, Mechanical properties of dried defatted spongy bone. *Acta Ort. Scand*;47:11-9.

Lotz, J.C., and Hayes, W.C., 1990, The Use of Quantitative Computed Tomography to Estimate Risk of Fracture of the Hip From Falls, *J. Bone Jt. Surg., Am.* 72:5, 689–700.

MacNeil J.A. and Boyd S.K., 2008, Bone strength at the distal radius can be estimated from high-resolution peripheral quantitative computed tomography and the finite element method. *Bone*. 42(6):1203-13.

Malik L., Stover M., Martin B. and Gibeling C., 2003, Equine cortical bone exhibits rising R-curve fracture mechanics, *J Biomech*, 36, 191–8.

Martin R.B, Burr D.R and Sharkey N.A, 1998. *Skeletal Tissue Mechanics.* Springer, New York.

McCalden, R.W., McGeough, J.A., Barker, M.B. and Court-Brown, C.M., 1993, Age-related changes in the tensile properties of cortical bone. *Journal of Bone and Joint Surgery* 75-A, 1193–1205.

McCalden RW, McGeough JA, Court-Brown CM. 1997, Age-related changes in the compressive strength of cancellous bone. The relative importance of changes in density and trabecular architecture. *J Bone Joint Surg Am*;79:421–7.

Morgan E.F. and Keaveny T.M., 2001, Dependence of Yield Strain of Human Trabecular Bone on Anatomic Site, *Journal of Biomechanics*, (34):569–577.

Morgan E.F., Yeh O.C., Chang WC and Keaveny T.M., 2001, Nonlinear behavior of trabecular bone at small strains, *J Biomech Eng*. 123(1):1-9.

Morgan E.F., Yeh O.C. and Keaveny T.M., 2005. Damage in trabecular bone at small strains, *Eur. J. Morphol*. 42 (1-2):13–21.

Mosekilde L, Mosekilde L., Danielsen C.C., 1987, Biomechanical competence of vertebral trabecular bone in relation to ash density and age in normal individuals. *Bone*;8:79-85.

Murakami, S. and Liu, Y., 1995. Mesh-dependence in local approach to creep fracture. *Int. J. Damage Mech*, 4, 230–250.

Nagaraja S., Couse T.L., Guldberg R.E., 2005, Trabecular bone microdamage and microstructural stresses under uniaxial compression, *Journal of Biomechanics* 38 707–716.

Nalla R.K., Kinney J.H., Ritchie R.O., 2003, Mechanistic fracture criteria for the failure of human cortical bone. *Nature Materials* 2, 164–168.

Nalla K., Kruzic J., Kinney H. and Ritchie O. 2005, Mechanistic aspects of fracture and R-curve behavior in human cortical bone, *Biomaterials*, 26, 217–31.

Natali A., Carniel E., Pavan P., 2008 Constitutive modelling of inelastic behaviour of cortical bone. *Med Eng Phys* 30(7):905–912.

Nazarian A., von Stechow D., Zurakowski D., Müller R. and Snyder B.D., 2008, Bone volume fraction explains the variation in strength and stiffness of cancellous bone affected by metastatic cancer and osteoporosis. *Calcif Tissue Int*. 83(6):368-79.

Niebur G.L., Feldstein M.J., Yuen J.C., Chen T.J. and Keaveny T.M., 2000, High-resolution finite element models with tissue strength asymmetry accurately predict failure of trabecular bone. *Journal of Biomechanics* 33, 1575–1583.

Parsamian G.P., 2002, Damage mechanics of human cortical bone. Phd Thesis, College of Engineering and Mineral Resources, West Virginia University.

Perilli E, Baleani M, Ohman C, Fognani R, Baruffaldi F, Viceconti M., 2008, Dependence of mechanical compressive strength on local variations in microarchitecture in cancellous bone of proximal human femur, *J Biomech*.;41(2):438-46.

Reilly D.T. and Burstein A.H., 1975, The elastic and ultimate properties of compact bone tissue, *J Biomech*. 8(6):393-405.

Rohl L., Larsen E., Linde F., Odgaard A., and Jorgensen J., 1991, Tensile and compressive properties of cancellous bone, *J. Biomech.*, 24, 1143–1149.

Saanouni K., Chaboche J.L. and Lesne P.M., 1989, On the creep crack-growth prediction by a non local damage formulation. *Eur. J. Mech. A/Solids*, 8, 437–459.

Sanyal A., Gupta A., Bayraktar H.H., Kwon R.Y., Keaveny T.M., 2012, Shear strength behavior of human trabecular bone. J Biomech.;45(15):2513-2519.

Seeman S., 2002, Pathogenesis of bone fragility in women and men, *The Lancet,* 1841–1850.

Smith C.B. and Smith D.A.,1976, Relations between age, mineral density and mechanical properties of human femoral compacta. *Acta Orthopaedica Scandinavica* 47, 496–502.

Stölken J.S, Kinney J.H., 2003, On the importance of geometric nonlinearity in finite-element simulations of trabecular bone failure. Bone 33:494–504.

Taylor D. and Lee T.C., 2003, A crack growth model for the simulation of fatigue in bone. *International Journal of Fatigue* 2, 387–395.

Thurner P.J., Erickson B., Jungmann R., Schriock Z., Weaver J.C., Fantner G.E., Schitter G., Morse D.E., Hansma P.K., 2007, High-speed photography of compressed human trabecular bone correlates whitening to microscopic damage. *Eng. Fract. Mech.* 74 (12), 1928–1941.

Tomar V., 2008, Modeling of dynamic fracture and damage in two-dimensional trabecular bone microstructures using the cohesive finite element method. *J Biomech Eng.* 130(2):021021.

Ural A. and Vashishth D., 2007, Anisotropy of age-related toughness loss in human cortical bone: A finite element study'. *Journal of Biomechanics*, 40, 1606–1614.

van Rietbergen B., Ulrich D., Pistoia W., Huiskes R., Ruegsegger P., 1998, Prediction of trabecular bone failure parameters using a tissue failure criterion. *Trans Orthop Res* Soc 23:550.

Vashishth D.J., Koontz Qiu S., Cannon-Lundin D., Yeni Y.N., Schaffler M.B. and Fyhrie D.P., 2000, In vivo diffuse damage in human trabecular bone. *Bone*. 26 (2):147–152.

Vashishth D. 2004, Rising crack–growth–resistance behavior in cortical bone: implications for toughness measurements. *J Biomech,* 37:943–6.

Verhulp E., Van Rietbergen B., Müller R. and Huiskes R., 2008, Micro-finite element simulation of trabecular-bone post-yield behaviour - effects of material model, element size and type, *Comput Methods Biomech Biomed Engin.*, (4):389-95.

Wang X., Zauel R. and Fyhrie D.P., 2008, Postfailure modulus strongly affects microcracking and mechanical property change in human iliac cancellous bone: A study using a 2D nonlinear finite element method, *Journal of Biomechanics* 41,2654– 2658.

Wolfram U., Wilke H.J., Zysset P.K., 2011, Damage accumulation in vertebral trabecular bone depends on loading mode and direction. *J Biomech.* 7, 44(6):1164-1169.

Yang D., Cox N., Nalla K. and Ritchie O., 2006, Re-evaluating the toughness of human cortical bone, *Bone* 38, 878–887.

Yeni Y., Brown C., Wang Z., Norman T., 1997, The influence of bone morphology on fracture toughness of the human femur and tibia. *Bone* 21:453–459.

Zioupos P., 1998, Recent developments in the study of failure of solid biomaterials and bone: fracture and pre-fracture toughness, *Mater. Science and Engng-C*, 6: 33-40.

In: Trabecular and Cortical Bone ISBN: 978-1-62808-766-6
Editor: V. Subramanium, M. Bhatnagar

Chapter IV

Application of Computational Modeling to Predict the Effect of Exercise on Whole Bone Strength

C. S. Florio
New Jersey Institute of Technology, Newark, NJ, US

Abstract

As bone adapts to changes in its environment, the loss of bone strength due to age, long term immobilization, or lengthy stays in reduced gravity conditions can lead to an increased risk of fracture. These strength adaptations are a result of phenomena that modify the mechanical properties of the bone. Changes occur both in the mineral content of the bone tissue and in the overall shape of the bone. Pharmaceutical countermeasures to decreased bone strength systemically alter the mechanisms that control the rates and amounts of mineral release from within bone tissue. This limits the reduction of the density of the bone and, thus, limits the weakening of the bone material itself. Rather than targeting the mineral loss mechanism directly, mechanical countermeasures, such as exercise, work independently of the density changes by amassing bone material on the surfaces of the cortical regions of loaded bones, thus altering the overall bone shape. When sufficient levels are reached, the increased size can improve the overall strength of the bone despite the material weakening that occurs with decreased density. While weight bearing exercises are widely accepted methods to offset the loss of bone strength with few adverse side-effects, quantitative comparisons of the benefits of the many different types of such exercises have not been well established through clinical or experimental means. Computational modeling methods have the ability to isolate the effects of individual system parameters, which is often difficult to achieve experimentally. Therefore, computational models may provide a means to objectively compare the benefits of the various suggested bone strengthening exercises. Developed models simulating cortical bone surface adaptation have shown the significant effect of loading parameters on bone shape, indicating that a better understanding of the relationships between the changes in the mechanical environment and the localized strength of a bone may be achieved through computational techniques. Therefore, the use of such computational models may be advantageous in the development of effective exercises for targeted strengthening of bone, particularly in regions that might be especially susceptible to fracture.

Introduction

Circumstances related to aging, prolonged immobilization due to injury, disease, or confinement, and even long term space travel all alter the mechanical loading environment of the musculoskeletal system. Such alterations have the potential to reduce muscle and bone strength. This reduced strength can lead to impaired functional mobility which can subsequently result in a reliance on assistance devices, a loss of independence, or even an increased risk of serious injury.

As a result, the mitigation of the loss of muscle and bone strength is of significant importance in the maintenance of normal levels of physical ability. Bone strength is often focused upon because bone fractures have the potential for severe consequences. As such, much work has been done in the development of methods to reduce such losses and improve bone strength. However, the current means of measuring the effectiveness of these countermeasures are limited. This has resulted in a wide range of recommendations with considerably varying outcomes. Therefore, the development of methods to quantitatively compare the capabilities of the many approaches to bone loss mitigation can be beneficial to the improvement of existing bone strengthening techniques.

Strength Adaptation of Bone

Bone has two main functions, that of structural support and that of a mineral reserve (see Figure 1). The first function, structural support, is influenced mainly by the size and shape of the bone. This is derived from the quantity and distribution of cortical bone tissue which comprises the overall structure and the visible surface of the bone [1]. The second function, that of a mineral reserve, is mainly a function of the cancellous bone tissue. This is evidenced by the more transient nature of the mineral content of the cancellous tissue which is located in the interior region of the bone [2]. While each type of bone tissue is suited to a particular role, both contribute to the mechanical function of the whole bone and both donate or store minerals, such as calcium and phosphorus, as needed by other physiological systems. Just like all other living tissues, from tree trunks [3] to lungs [4], bone has the ability to adjust to changes in environmental conditions. This allows it to maintain a relatively steady level of performance while operating efficiently within the whole body system. Such adjustments are often referred to as functional adaptations [5]. In bone, these adaptations have a direct impact on strength. Usually, the functional adaptations of the bone create a structure with suitable strength to operate appropriately under loading conditions that are typically encountered. However, sometimes, these adaptations create a bone structure too weak to withstand everyday loads, resulting in an increased likelihood of fracture. In these situations, the detrimental effects of aging, long term immobilization, or lengthy stays in reduced gravity environments can be mitigated by taking advantage of the bone's strength adaptation mechanisms.

The ability of bone to modify its strength due to changes in its local mechanical environment occurs through two independent processes (see Figure 1). As in mechanical design, an object's strength can be adjusted through material choice (intrinsic properties) or shape (extrinsic properties). The strength modifications of bone target both of these features. The first strength modification process takes advantage of bone's role as a mineral reservoir to store or release minerals as needed. It also functions to repair any local damage to the bone structure. The effects of this process result in changes to both the mineral content of the bone tissue and to its porosity. Therefore, this first process affects the intrinsic mechanical properties of the bone, namely, the density of the bone tissue material [6]. The function of this type of strength adaptation can be easily tracked through measurements of bone density, or, more specifically, bone mineral density (BMD).

This is also sometimes referred to as “bone mass”. While more prevalent in the cancellous bone tissue, changes to the density of the cortical bone also occur through this phenomenon [7].

The second strength modification process works by altering the size and shape of the bone itself. This process occurs through the deposit or removal of material from the surface of the bone, thus changing the bone’s extrinsic properties. By changing the shape of the bone, its overall strength is also modified [6]. These changes occur both on the exterior (periosteal) and interior (endosteal) surfaces of cortical bone. While not as broadly recognized as bone mineral density as a measure of bone strength, bone size and shape play a significant role in whole bone strength. For instance, it is widely noted that the endosteal and periosteal diameter of a bone, from a femur to a finger, will increase in response to the natural decrease of the bone’s density with age [1, 2, 8]. Another example of bone shape change in response to need for increased bone strength is the development of a fracture callus, a large bulbous structure that quickly grows as a kind of support patch surrounding a bone fracture. While made of material that is weaker than typical cortical bone tissue, the increased profile at the fracture site provides quick stability as well as strength to the damaged area as the bone heals [9, 10]. A similar structure to the fracture callus is a bone spur. This spur is a boney growth that sometimes develops near an arthritic joint in an attempt to reduce stresses at the joint by increasing the contact area [1]. Additionally, some studies have shown that people with a higher risk of stress fractures may have smaller sized, not less dense, bones [11-14].

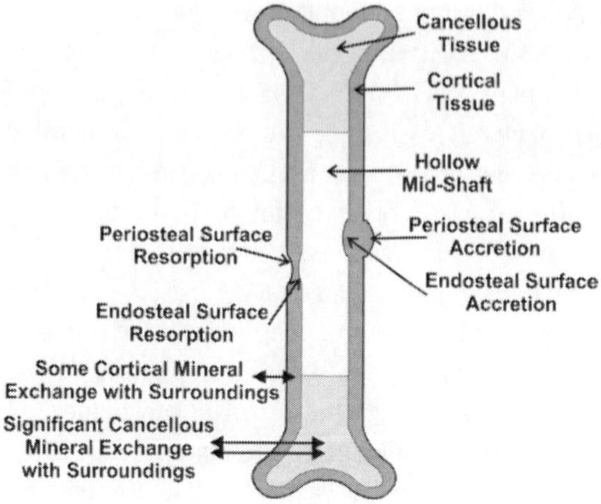

Figure 1. Typical bone structure and strength adaptation mechanisms.

Surface resorption and accretion of cortical bone occur locally and independently from each other and from the changes in cortical and cancellous bone density [9]. Under small fluctuations in the mechanical environment, these phenomena work to produce small net changes in bone strength. However, under certain conditions, the rates or amounts of change to the bone density or shape can cause a net loss or gain of bone strength. As a result of processes related to aging or to decreased environmental loading, significant loss of bone strength can occur. Environmental loading can include body forces due to gravity, external forces on the whole body system, such as applied weights or impact forces, or forces internal to the body system, such as those created by muscles. An adaptive response to a reduced external loading environment may create weaker bone which is well suited for the new environment's typically smaller magnitude loads. However, should larger environmental forces resume, such as when rehabilitating from prolonged bedrest or arriving to a planet with higher gravitational acceleration after long term space travel, the potential for fracture can increase. Age related loss of bone strength can occur without changes to external environmental loads and is often due to decreased material properties. Regardless of the cause, the level of strength reduction can be lessened, and the strength may even be increased, by modifying the rates and amounts of bone shape and density changes through the use of various interventions and countermeasures.

Current Countermeasures

Currently, there are two main types of methods to counter the loss of bone strength. Each targets a different bone strength modification process discussed above. The first is pharmaceutical based. Pharmaceutical based countermeasures target changes in bone density. They alter the rates of mineral fluctuation from the storage reservoirs of bone. Additionally, they slow the repair processes in the bone, thus reducing the amounts and rates of loss of minerals and of increase of material porosity [7]. This leaves a denser, stronger material. However, there have been some concerns with potentially unwanted side effects including the increased risk of fracture in atypical locations [15]. There are various theories as to the cause of these fractures. One theory is that the reduction in the rate of repair that accompanies the reduction in the rate of loss leads to larger defects in the bones that are more likely to propagate across the whole bone structure under repeated loading [16, 17]. Another theory is that the increased mineralization makes a stiffer, more brittle material [18]. Despite the risks, pharmaceutical based methods can contribute to a reduced potential of fracture under typical loads for those with severe weakening of the bone. Additionally, since bone density is readily measured, the effectiveness of pharmaceutical based countermeasures is easy to quantify and track.

The second type of countermeasure is through mechanical based methods, mainly exercise. Because the main function of many bones is to allow movement of the body through their rotation about joints, forces due to the pull of the muscles on bones create some of the largest stresses acting on bones [2, 19, 20]. Therefore, altering the muscle forces is a significant way to alter the bone's mechanical environment and to induce bone strength adaptations. Unlike pharmaceutical based countermeasures that work to prevent reductions in bone strength by preventing loss of bone material, exercise based countermeasures work to improve bone strength by stimulating bone formation [16]. Quantifying the effectiveness of exercise based countermeasures is usually more difficult than measuring the effectiveness of pharmaceutical based ones [21].

The relationship between changes in muscle strength and changes in bone strength has been clearly shown both clinically and through animal experiments. Much of this has been through studies where subjects were immobilized either through cast, surgery, or bed confinement [9, 22-26]. Measures of bone strength, such as geometry, density, or blood or urine mineral content, are tracked along with measures of muscle strength, such as muscle size. Controlled increased loads are applied, either internally through exercise or externally through mechanisms that simulate the resultant effects of muscle function, and changes in the bone and muscle strength measures are determined. Similar studies have also been undertaken in microgravity. Bone and muscle changes in astronauts during their time in space and upon return to Earth have been tracked from Gemini and Soyuz missions to the current International Space Station, where the time spent in microgravity represents the "immobilization period" [27-31]. Over a wide range of subjects, from hamsters [32] to turkeys [33], sheep [24] to dogs [34, 35], and pigs [36] to humans [37], the general observations are consistent. Reduced loading results in a relatively rapid decrease in bone mineral density and porosity, especially of the cancellous bone tissue, which then levels off to a new equilibrium state. Muscle size, and thus strength, also decreases rather rapidly. Prolonged reduced loading eventually results in an overall decrease in the size (volume) of the cortical bone. Upon resuming original loads, the process is reversed, and changes in bone size lag behind changes in muscle size and bone mineral content. Prolonged increased loading has demonstrated the ability to significantly increase cortical bone thickness and overall bone size, even after changes in bone density have stopped.

While bone is able to adapt both its material (intrinsic) and geometric (extrinsic) properties, alterations in bone geometry have been shown to induce the most significant effects on overall bone strength [38]. The changes in bone density are typically highly transient and occur mainly in the cancellous tissue. Cortical bone is generally an order of magnitude stronger than cancellous bone. Thus, even significant changes in cancellous bone density do not considerably affect the overall strength of the bone [1]. It has been found that the material properties of bone actually vary little with age and even species compared to differences in whole bone strength due to bone shape [39]. Because whole bone strength is mainly a function of the cortical bone, and cortical bone geometry changes are usually greater than changes in cortical bone density, it has been suggested that osteoporotic fracture prevention methods might benefit from a research focus on altering the bone's surface geometry rather than its mineral content [21, 40]. Additionally, the methods that result in these extrinsic changes, often mechanical exercise interventions, may provide benefits to other physiological systems in addition to the bone, while the bone is the sole target of pharmaceutical countermeasures.

Numerous studies have been performed to quantify the effects of exercise on changes in bone strength. They have focused on many different types of high and low intensity exercises including walking, running, cycling, swimming, jumping, soccer, tennis, squash, and various forms of weightlifting [41-45]. Some studies have instituted a prescribed exercise regimen where bone mineral density or bone size was tracked as the exercise program was followed. Trends in these measures were compared to trends in the same measures for a control group that did not exercise. Others studies have been retrospective, where the measures of bone strength of athletes who specialize in certain activities were compared. Many of these studies have included mixed loading modes. Most studies have encountered prevalent issues with protocol compliance and intersubject variability. Generally, regardless of the loading mode, most exercises have shown the potential to mitigate the loss of bone strength. However, except for general conclusions that high force or high impact loads like resistance training seem to have a greater effect on bone strength than low impact exercise modes like walking or swimming, these studies have not been able to determine specific exercise modes that are most beneficial in improving bone strength.

Both experimental methods on animal subjects and human clinical studies have not been successful in isolating the effects of loading mode on bone strength [29]. As such, current exercise protocols offer a wide array of techniques leading to significant variations in effectiveness [46]. Computational studies are often employed in mechanical analysis to gain insight into phenomena that are difficult to measure, isolate, or see through experimental means. Therefore, computational methods may be better suited to isolate specific exercise parameters that have the potential to induce the greatest bone strength adaptations without the variability and restrictions that limit experimental and clinical investigations.

Computational Modeling

Most computational models of the changes to a bone's geometry as a result of changes to its mechanical environment use mathematical optimization methods coupled with the numerical solution of the response of the bone studied to a prescribed set of applied boundary conditions (loads and constraints). In the application of these models, the bone geometry is broken into a number of small regions, finite elements or boundary elements, where the governing equations can more easily be solved to determine the behavior of the bone in response to the particular set of external forces or boundary conditions applied. This response can be represented through various measures of the internal mechanical state of the bone such as deformation, strain, stress, or energy stored in the bone system due to the deformation, referred to as strain energy. The measures are determined at discrete locations within the bone volume, or nodes, so that the variation of the mechanical response throughout the bone system can be determined. Based on this distribution of the local mechanical state of bone, the local material accretion or resorption is determined using a numerical model of the bone's strength adaptation phenomena. The volume and shape changes of the bone are then simulated by moving the location of discrete points, usually element nodes or calculated shape control points, on the boundary surface of the bone model. Using such a method, the governing conservation equations of the system are solved quasi-statically, and the shape optimization progresses iteratively. In this way, the mass of the system remains constant while the mechanical state of the bone is being determined, eliminating the need to include conservation of mass requirements in the governing equations.

As the computational solution proceeds, the shape of the bone is altered at each iteration, driven by the adaptive model, through the progress of the chosen measure of the bone's local mechanical state toward a specified optimization goal. With the new shape at each iteration, the boundary conditions are again imposed, the governing equations solved, and the new mechanical state of the bone determined. From this new distribution of the deformations and stresses within the bone, the strength adaptation model is again employed and the boundary surfaces again moved. This process is repeated until a defined stopping criteria is reached. This stopping criteria is usually a measure of convergence or the achievement of the stated optimization goal. The stopping criteria can also be the achievement of a physical size limit of the bone system, such as maximum or minimum bone width or cortical thickness. During the iterative shape changing process, remeshing or mesh smoothing methods are often employed to counter the distortions to the discretization of the bone volume (the mesh) that result from moving the boundary surfaces and, therefore, stretching the elements attached to these surfaces. This remeshing or smoothing ensures the validity of the finite element or boundary element model during the alterations of the system's shape [47]. In models using finite element methods, smoothing or remeshing is applied both on the external surface and within the internal volume of the geometry. The whole procedure can be applied to three-dimensional models as described above or to two-dimensional models where the boundary surfaces become boundary curves and internal volumes become internal areas.

The techniques that are used for the computational modeling of the shape changes that occur due to the cortical bone's strength adaptation phenomenon as a result of local changes in the loading environment are very similar to the shape optimization processes used in the design of inert mechanical components based on size and strength design criteria. In both bone strength adaptation simulations and shape optimization in mechanical design, many different types of mathematical optimization methods, from gradient based to gradientless, and local to global, have been implemented [48-53]. Additionally, various measures of the local mechanical state have been used to drive the changes to the locations of the discrete points along the surfaces of the shapes being optimized. For example, average, equivalent, or principal stresses or strains, individual components of the stress or strain tensors, and energy measures, such as strain energy density, as well as rates and spatial gradients of these quantities have all been considered in the development of various bone shape adaptation and mechanical shape optimization models

[54-57]. Whether applied to inert or biological materials, the specific optimization goal chosen creates a progression towards a more uniform distribution of the local mechanical measure, thus reducing the maximum surface stresses and strains. Because of the geometric complexity and resulting large number of design points, whether they be surface nodes or other such control points, gradientless optimization methods in particular are often employed in the study of bone adaptation. In these methods, a threshold is usually specified, below which bone is removed and the volume or area is reduced, while above which material is added and bone volume increases. This threshold may be a constant value determined based on experimental measures [58] or chosen based on mechanical properties, such as yield strength. It can vary with location [9], magnitude of applied loads [59], or even with iteration, such as with an average measure or similar quantity calculated from the current mechanical state of the bone [60]. The threshold may also be a range of values between which no changes occur [61]. These gradientless methods are less mathematically rigorous and do not result in a global optimal solution. However, they are suitable for shape optimizations problems which typically are solved over a limited dimensional range.

Despite the large array of methods employed, most computational models developed have successfully predicted similar trends in the changes of the bone shape in response to specified applied external loads when compared to trends that have been observed experimentally for similar load sets. Additionally, computational models that have isolated the basic loading modes of compression, bending, and torsion have revealed similar trends to each other for each basic loading mode despite significant differences in the specific modeling methods employed [62-64]. An illustrative case is that of an initially circular hollow cylinder, which is a good representation of the diaphyseal region of a long bone. When the basic loading modes are applied, some specific and repeatable types of shape changes result. Figure 2 shows the typical changes, which are described as follows. Depending on the relative magnitudes of the optimization goal, or threshold, and the applied load, an axially directed compressive force (Figure 2a) would create a uniform stress distribution across an axial, or transverse, cross section of the long bone. This would result in a uniform increase or decrease in the inner and outer diameters, representing either uniform accretion or resorption of bone material at the same rate on both inner and outer cortical surfaces. A torque about the longitudinal axis of the long bone (Figure 2b) would cause the inner and outer surfaces to both grow, both decay, or the outer grow and the inner decay. The levels of accretion or absorption on the outer surfaces would be higher than on the inner surface since the torsional stresses/strains, the drivers for growth or decay, are maximum on the outer surface of an object. A bending moment about an axis normal to the axial direction of the long bone (Figure 2c) would create an egg shaped geometry for each transverse cross section of the bone at both the inner and outer boundaries. Under such loading, decay would occur along the neutral bending axis and growth would occur 90 degrees from this axis along the region of high stress. The levels of growth and decay at each inner and outer boundary would depend on the relative values of the applied load and optimization goal. Combinations of these basic loading modes would make predictable combinations of these described geometries. By isolating the basic loading modes though computational simulations, it is then possible to create desirable shape changes through a controlled application of combinations of these modes. These desired shape changes can be based on bone strength requirements. Such knowledge can be employed when creating the complex combinations of loads used in bone strengthening exercise routines.

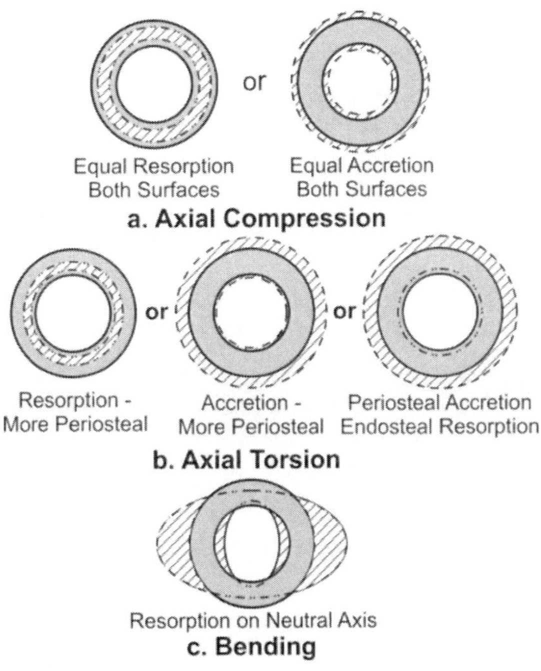

Figure 2. Effect of loading mode on cross-sectional geometry of mid-shaft cortical bone.

While the potential for controlled bone strength changes through the utilization of the conclusions drawn from computational simulations is apparent, improvements to existing modeling capabilities are necessary before such models can be widely applied. Many of the current models that predict the adaptive shape changes of the cortical bone are limited in either the representation of the structure of the bone or its adaptive response. In order to exemplify the effects of boundary conditions on shape changes, the bone's geometry and material properties are often simplified [53, 65, 66]. Such simplifications are also made in order to reduce the time required to complete the simulation, as the shape optimization models can be computationally intensive for the large number of surface nodes needed to depict a complex geometry. Often a two-dimensional model of a simple cross-section of a bone is used [62, 63]. While these two-dimensional models appropriately reflect the relative effects of the loads on the bones, the simplifications ignore the effects of axial variations. This may limit the accuracy of the shape predictions. Likewise, some simulations may use more complex three-dimensional geometries and material models that include both cortical and cancellous bone tissues [67]. These may give a better representation of the mechanical state of the bone. However, if model parameters are chosen to match a set of arbitrary experimental data, the models cannot be more generally applied to subsequent studies. Additionally, convergence and comparison criteria that can be applicable across many different sets of boundary conditions, such as external forces, and sets of system conditions, such as the bones modeled or the limb orientation, have not yet been established [47, 68-71].

Even the more realistic bone models used in current shape adaptation simulations are not as complex as those currently used to simply determine the response of the bone to mechanical loading without the predication of the ensuing strength adaptation. Such models used to predict the state of the internal stresses and strains in a bone often include more detailed geometry, material models, and boundary conditions. These complex models often include multi-bone systems, such as all the bones that comprise a limb or section of the spine [72, 73]. They often use more realistic bone geometry, sometimes obtained from bone scans [74]. The effects of the contact and interactions of the bones with soft tissues such as cartilage and ligaments are often included [75]. Additionally, the application of the individual muscle forces to the boundaries of the bones can be simulated [76]. Such models have shown that the more accurate the representation of the physiological conditions, the better the reproduction of experimental results. Studies have found that model predictions are strongly influenced by the model's geometric complexity and the representation of system boundary conditions. However, the results are more sensitive to the boundary conditions used [77]. Thus, the careful consideration of the applied boundary conditions as well as the geometric modeling features in shape adaptation simulations would likely prove beneficial.

The need for improving the representation of physiological conditions in computational models is of particular importance when considering the muscle forces acting on the bone's surface. This is because muscles produce some of the largest forces applied to bones, often greater than those from forces at the joints or those applied externally to the whole body. Studies have found that the method used to represent the application and distribution of muscle forces over the bone's surface significantly affects the resulting stress and strain fields in the bone [78]. Models simulating the muscle-tendon complex both in function and size of load application area have been shown to produce more gradual and realistic transitions between regions of high and low stress and strain [76]. Additionally, models varying the number of muscles included in the model have shown that the more complete the physiological model of the muscle loading, meaning simulating the individual muscles rather than just functional groups, the more conservative the estimates of the resulting stress/strain distributions since smaller local variations occur as the same resultant force is distributed over more active muscles [77, 79]. Thus, an important part of modeling the mechanical environment of a bone and its resulting shape changes is an appropriate portrayal of the function of the individual active muscles.

The incorporation of an improved representation of the muscles acting on the bone surfaces involves the accurate determination of the magnitudes of the produced muscle forces. Usually a resultant force generated by a system of muscles is known, such as the force measured between the foot and a fixed plate. The muscle activity in the leg that creates this resultant force must be calculated. Many methods to determine the individual muscle forces that produce a known resultant force have been developed. When static conditions prevail, such as occur during activities where a prescribed force is generated as in resistance exercises, inverse kinematics methods can be employed. Often, mathematical optimization methods are used to solve these inverse kinematics problems. Unlike the shape optimization problems described earlier, gradient based techniques are usually required so that global optima, a unique set of muscle forces, can be obtained. Applications of such models have been able to show that even relatively small changes in the overall resultant force are achieved through significant variations in the generated muscle forces. For example, the active muscles in the leg used to create a resultant force during an isometric exercise vary greatly with the direction of the resultant force [80]. Additionally, for the same magnitude of this resultant force, the value of the force generated by a given muscle that contributes to the resultant force can vary up to two orders of magnitude depending on the resultant's direction [81]. Applying these calculated muscle forces to a finite element model of the leg system can help improve the understanding of the local variations in stress and strain within a bone for different load-bearing activities. Such models can help identify regions under high stress that can benefit from strengthening. For example, as shown in Figure 3, a finite element model of a leg system with directly applied muscle forces shows significantly different stress distributions and up to an order of magnitude difference in maximum stress values solely from changing the direction, not the magnitude, of the resultant force between the foot and a fixed plate as a direct result of the differences in muscle activity [82]. Similarly, the variations in bone stresses due to variations in muscle activity of altered gait patterns from neuromuscular disease, damage or simply age can be determined by incorporating accurate modeling of muscle forces [83]. Considerable insight into the effects of generated muscle forces on bone strength can, therefore, be achieved by including bone strength adaptation prediction modeling into these more anatomically complex musculoskeletal models.

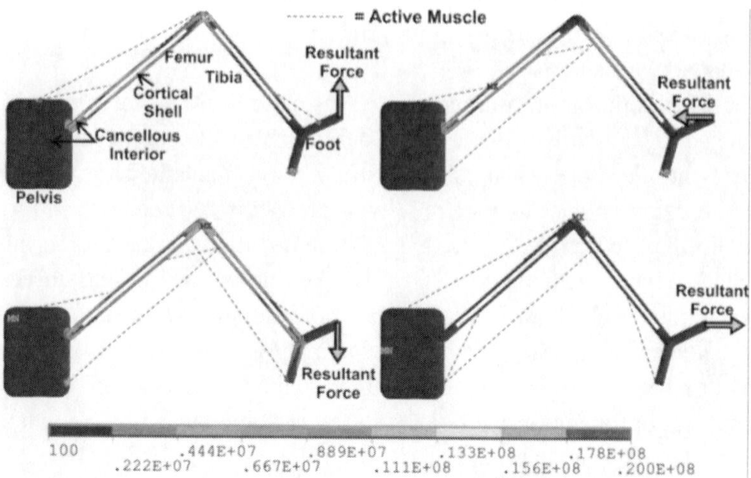

Figure 3. Variation in active muscles and von Mises stress distributions in a transverse cross section of a leg bone system for various directions of a constant magnitude resultant force.

Conclusion

A reduction of bone strength losses due to age or immobilization through mechanical means such as exercise has been demonstrated both experimentally and clinically. However, clinical studies and animal experiments have not been successful in determining the relative effectiveness of different exercise regimens. The application of computational models to study bone growth phenomena can yield new insight not capable of being captured through experimental studies of living subjects. Methods have been developed to predict the local shape changes of cortical bone that occur as a result of applied mechanical loads. However, these simulations have often been limited in their complexity and universality. Separately, methods have been developed to create more accurate representations of bone systems, especially though the addition of the effects of individual muscles that together generate a desired action or resultant force. Thus, the inclusion of these more accurate representations of musculoskeletal systems into the shape and strength adaptation models along with the appropriate selection of comparison measures have the potential to provide a quantitative means of determining the relative effectiveness of an exercise at promoting bone growth without the limitations of standard experimental means.

Such coupled models, which are currently in development, can be used to calculate bone shape changes for any given set of system conditions. For example, simulations could be run for various sets of resultant forces and limb orientations. Localized bone strength changes could then be compared to determine the system conditions that produce enhancements to bone strength that most effectively resist the types of loads that are typically encountered in daily activities. Resultant forces and limb orientations can be readily measured experimentally through standard motion analysis measurement equipment. These design inputs are, therefore, easily incorporated both into the computational simulations and into the clinical development of exercise regimens. Thus, computational modeling methods can be used to isolate the effects that changes in system condition parameters have on the distribution of the local mechanical state of a bone, the ensuing shape changes, and the resulting effects on localized bone strength. Such simulations can enhance the study of the mechanical control of the localized functional strength adaptation of the cortical bone. This knowledge can assist in the development of exercise interventions that target regions of bone that may be especially susceptible to fracture. Therefore, the development and application of computational models of cortical bone strength adaptations can benefit not only the understanding of the form and function of cortical bone, but also its clinical management.

References

[1] Martin, RB; Burr, D. *Structure, Function and Adaptation of Compact Bone*. New York: Raven Press; 1989.

[2] Currey, JD. *Bones: Structure and Mechanics*. Princeton, NJ: Princeton University Press; 2002.

[3] Archer, RR; Wilson, BF. Mechanics of the Compression Wood Response. *Plant Physiol*. 1970; 46, 550-6.

[4] Cowan, MJ; Crystal, RG. Lung growth after unilateral pneumonectomy: Quantitation of collagen synthesis and content. *Am Rev Respir Disease*. 1975; 111, 267-76.

[5] Roesler, H. Some historical remarks on the theory of cancellous bone structure. In: Cowin SC, editor. *Mechanical Properties of Bone* at the Joint ASME-ASCE Applied Mechanics, Fluids Engineering and Bioengineering Conference, Boulder, Colorado June 22-24, 1981; New York: American Society of Mechanical Engineers; 1981, 27-42.

[6] Frost, HM. Strain and other mechanical influences on bone strength and maintenance. *Curr Opin Orthop*. 1997; 8(5), 60-70.

[7] Martin, RB; Burr, DB; Sharkey, NA. *Skeletal Tissue Mechanics*. New York: Springer; 1998.

[8] Dequeker, J. Periosteal and endosteal surface remodeling in pathologic conditions. *Invest Radiol*. 1971; 6(4), 260-5.

[9] Cowin, SC, editor. *Bone Mechanics Handbook*. Second Edition. New York: Informa Healthcare; 2008.

[10] Garcia-Aznar, JM; Kuiper, JH; Gomez-Benito, MJ; Doblare, M; Richardson, JB. Computational simulation of fracture healing: Influence of interfragmentary movement on the callus growth. *J Biomech*. 2007; 40, 1467-76.

[11] Milgrom, C; Giladi, M; Simkin, A; Rand, N; Kedem, R; Kashtan, H; et al. The Area Moment of Inertia of the Tibia: A Risk Ractor for Stress Fractures. *J Biomech*. 1989; 22(11/12), 1243-8.

[12] Beck, TJ; Ruff, CB; Mourtada, FA; Shaffer, RA; Maxwell-Williams, K; Kao, GL; et al. Dual-Energy X-Ray Absorptiometry Derived Structural

Geometry for Stress Fracture Prediction in Male U.S. Marine Corp Recruits. *J Bone Miner Res*. 1996; 11(5), 645-53.

[13] Ackerman, KE; Putman, M; Guereca, G; Taylor, AP; Pierce, L; Herzog, DB; et al. Cortical microstructure and estimated bone strength in young amenorrheic athletes, eumenorrheic athletes and non-athletes. *Bone*. 2012; 51(4), 680-7.

[14] Crossley, K; Bennell, KL; Wrigley, T; Oakes, BW. Ground reaction forces, bone characteristics, and tibial stress fracture in male runners. *Med Sci Sports Exerc*. 1999; 31(8), 1088-93.

[15] Eastell, R; Walsh, JS; Watts, NB; Siris, E. Bisphosphonates for Postmenopausal Osteoporosis. *Bone*. 2011; 49, 82-8.

[16] Hawkey, A. Low magnitude, high frequency signals could reduce bone loss during spaceflight. *JBIS - Journal of the British Interplanetary Society*. 2007; 60(8), 278-84.

[17] Turner, CH. Homeostatic control of bone structure: An application of feedback theory. *Bone*. 1991; 12(3), 203-17.

[18] Akkus, O; Polyakova-Akkus, A; Adar, F; Schaffler, MB. Aging of Microstructural Compartments in Human Compact Bone. *J Bone Miner Res*. 2003; 18(6), 1012-9.

[19] Burr, DB. Muscle strength, bone mass, and age-related bone loss. *Journal of Bone and Mineral Research: The Official Journal of the American Society for Bone and Mineral Research*. 1997; 12(10), 1547-51.

[20] Pavy-Le Traon, A; Heer, M; Narici, MV; Rittweger, J; Vernikos, J. From space to Earth: advances in human physiology from 20 years of bed rest studies (1986-2006). *Eur J Appl Physiol*. 2007; 101, 143-97.

[21] Frost, HM. A 2003 update of bone physiology and Wolff's law for clinicians. *Angle Orthod*. 2004; 74(1), 3-15.

[22] Deitrick, JE; Whedon, GD; Shorr, E. Effects of immobilization upon various metabolic and physiologic functions of normal men. *The American Journal of Medicine*. 1948; 4(1), 3-36.

[23] Jaworski, ZFG; Liskova-Kiar, M; Uhthoff, HK. Effect of long term immobilization on the pattern of bone loss in older dogs. *Journal of Bone and Joint Surgery - Series B*. 1980; 62B(1), 104-10.

[24] Lanyon, LE; Goodship, AE; Pye, CJ; MacFie, JH. Mechanically adaptive bone remodelling. *J Biomech*. 1982; 15(3), 141-54.

[25] Coutts, RE; Akeson, WH; Woo, SLY. Comparison of stainless steel and composite plates in the healing of diaphyseal osteotomies of the dog

radius: report on a short term study. *Orthop Clin North Am*. 1976; 7(1), 223-9.

[26] Steinberg, ME; Trueta, J. Effects of activity on bone growth and development in the rat. *Clin Orthop Relat Res*. 1981; No. 156, 52-60.

[27] Evans, CA; Robinson, JA; Tate-Brown, J; Thumm, T; Crespo-Richey, J; Baumann, D; et al. International Space Station Science Research Accomplishments During the Assembly Years: An Analysis of Results from 2000-2008. Houston, TX: NASA, Center JS; 2009 NASA/TP-2009-213146-Revision A.

[28] LeBlanc, AD; Spector, ER; Evans, HJ; Sibonga, JD. Skeletal responses to space flight and the bed rest analog: A review. *Journal of Musculoskeletal Neuronal Interactions*. 2007; 7(1), 33-47.

[29] Sibonga, JD; Cavanagh, PR; Lang, TF; LeBlanc, AD; Schneider, VS; Shackelford, LC; et al. Adaptation of the skeletal system during long-duration spaceflight. *Clin Rev Bone Miner Metab*. 2007; 5(4), 249-61.

[30] Vico, L; Collet, P; Guignandon, A; Lafage-Proust, MH; Thomas, T; Rehailia, M; et al. Effects of long-term microgravity exposure on cancellous and cortical weight-bearing bones of cosmonauts. *Lancet*. 2000; 355(9215), 1607-11.

[31] Schneider, V; Oganov, V; LeBlanc, A; Rakmonov, A; Taggart, L; Bakulin, A; et al. Bone and body mass changes during space flight. *Acta Astronaut*. 1995; 36(8-12), 463-6.

[32] Borer, KT; Kuhns, LR. Radiographic evidence for acceleration of skeletal growth in adult hamsters by exercise. *Growth*. 1977; 41(1), 1-13.

[33] Lanyon, LE; Rubin, CT. Static Vs Dynamic Loads as an Invluence on Bone Remodelling. *J Biomech*. 1984; 17(12), 897-905.

[34] Uhthoff, HK; Jaworski, ZFG. Bone loss in response to long-term immobilisation. *Journal of Bone and Joint Surgery - Series B*. 1978; 60B, 420-9.

[35] Tonino, AJ; Davidson, CL; Klopper, PJ; Linclau, LA. Protection from stress in bone and its effects: experiments with stainless steel and plastic plates in dogs. *Journal of Bone and Joint Surgery - Series B*. 1976; 58B(1), 107-13.

[36] Woo, SL-Y; Kuei, SC; Amiel, D; Gomez, MA; Hayes, WC; White, FC; et al. The effect of prolonged physical training on the properties of long bone: a study of Wolff's law. *Journal of Bone and Joint Surgery - Series A*. 1981; 63A(5), 780-7.

[37] Spector, ER; Smith, SM; Sibonga, JD. Skeletal Effects of Long-Duration Head-Down Bed Rest. *Aviat Space Environ Med.* 2009; 80(5), A23-A8.

[38] Roesler, H. The history of some fundamental concepts in bone biomechanics. *J Biomech.* 1987; 20(11/12), 1025-34.

[39] Frost, HM. From Wolff's Law to the Utah Paradigm: Insights about bone physiology and its clinical applications. *The Anatomical Record.* 2001; 262, 398-419.

[40] Ferretti, JL; Frost, HM; Gasser, JA; High, WB; Jee, WSS; Jerome, C; et al. Perspectives on osteoporosis research: Its focus and some insights from a new paradigm. *Calcif Tissue Int.* 1995; 57(6), 399-404.

[41] Warner, SE; Shea, JE; Miller, SC; Shaw, JM. Adaptations in cortical and trabecular bone in response to mechanical loading with and without weight bearing. *Calcif Tissue Int.* 2006; 79(6), 395-403.

[42] Engelke, K; Kemmler, W; Lauber, D; Beeskow, C; Pintag, R; Kalender, WA. Exercise maintains bone density at spine and hip. EFOPS: A 3-year longitudinal study in early postmenopausal women. *Osteoporosis Int.* 2006; 17(1), 133-42.

[43] Vainionpaa, A; Korpelainen, R; Sievanen, H; Vihriala, E; Leppaluoto, J; Jamsa, T. Effect of impact exercise and its intensity on bone geometry at weight-bearing tibia and femur. *Bone.* 2007; 40, 604-11.

[44] Nikander, R; Kannus, P; Dastidar, P; Hannula, M; Harrison, L; Cervinka, T; et al. Targeted exercises against hip fragility. *Osteoporosis Int.* 2009; 20(8), 1321-8.

[45] Guadalupe-Grau, A; Fuentes, T; Guerra, B; Calbet, JAL. Exercise and bone mass in adults. *Sports Med.* 2009; 39(6), 439-68.

[46] Ruff, C; Holt, B; Trinkaus, E. Who's afraid of the big bad Wolff?: "Wolff's Law" and bone functional adaptation. *Am J Phys Anthropol.* 2006; 129, 484-98.

[47] Schnack, E. Optimization Procedure for Stress Concentrations by the Finite Element Technique. *International Journal for Numerical Methods in Engineering.* 1979; 14(1), 115-24.

[48] Haftka, RT; Gurdal, Z; Kamat, MP. *Elements of Structural Optimization.* Boston: Kluwer Academic Publishers; 1990.

[49] Ding, Y. Shape Optimization of Structures: A Literature Survey. *Computers and Structures.* 1986; 24(6), 985-1004.

[50] Imam, MH. Three-Dimensional Shape Optimization. *International Journal for Numerical Methods in Engineering.* 1982; 18(5), 661-73.

[51] Sonmez, FO. Shape Optimization of 2D structures using simulated annealing. *Computer Methods in Applied Mechanics and Engineering*. 2007; 196(35-36), 3278-99.

[52] Mattheck, C; Baumgartner, A; Walther, F; editors. *Optimization Procedures by Use of the Finite Element Method*. American Society of Mechanical Engineers, Petroleum Division (Publication) PD; 1994; London, Engl: Publ by ASME.

[53] Chen, G; Pettet, GJ; Pearcy, M; McElwain, DLS. Modelling external bone adaptation using evolutionary structural optimisation. *Biomechanics and Modeling in Mechanobiology*. 2007; 6(4), 275-85.

[54] Carter, DR. Mechanical loading histories and cortical bone remodeling. *Calcif Tissue Int*. 1984; 36(Suppl. 1), S19-S24.

[55] Cowin, SC; Van Buskirk, WC. Surface bone remodeling induced by a medullary pin. *J Biomech*. 1979; 12(4), 269-76.

[56] Brown, TD; Pedersen, DR; Gray, ML; Brand, RA; Rubin, CT. Toward an identification of mechanical parameters initiating periosteal remodeling: A combined experimental and analytic approach. *J Biomech*. 1990; 23(9), 893-7.

[57] Judex, S; Gross, TS; Bray, RC; Zernicke, RF. Adaptation of bone to physiological stimuli. *J Biomech*. 1997; 30(5), 421-9.

[58] Burr, DB; Martin, RB; Schaffler, MB; Radin, EL. Bone remodelling in response to in vivo fatigue microdamage. *J Biomech*. 1985; 18(3), 189-200.

[59] Beaupre, GS; Orr, TE; Carter, DR. An approach for time-dependent bone modeling and remodeling-theoretical development. *J Orth Res*. 1990; 8(5), 651-61.

[60] Le, C; Bruns, T; Tortorelli, D. A Gradient-Based, Parameter-Free Approach to Shape Optimization. *Computer Methods in Applied Mechanics and Engineering*. 2011; 200, 985-96.

[61] Carter, DR; van der Meulen, MCH; Beaupre, GS. Mechanical factors in bone growth and development. *Bone*. 1996; 18(1), 5S-10S.

[62] Mittlmeier, T; Mattheck, C; Dietrich, F. Effects of mechanical loading on the profile of human femoral diaphyseal geometry. *Med Eng Phys*. 1994; 16(1), 75-81.

[63] Levenston, ME; Beaupre, GS; Carter, DR. Loading Mode Interactions in Simulations of Long Bone Cross-Sectional Adaptation. *Comput Methods Biomech Biomed Eng*. 1998; 1(4), 303-19.

[64] Florio, CS; Narh, KA. Effect of modeling method on prediction of cortical bone strength adaptation under various loading conditions. *Meccanica*. 2012.

[65] Cowin, SC; Hart, RT; Balser, JR; Kohn, DH. Functional adaptation in long bones: Establishing in vivo values for surface remodeling rate coefficients. *J Biomech*. 1985; 18(9), 665-71.

[66] Hart, RT; Davy, DT; Heiple, KG. A computational method for stress analysis of adaptive elastic materials with a view toward applications in strain-induced bone remodeling. *J Biomech Eng*. 1984; 106(4), 342-50.

[67] Chennimalai Kumar, N; Dantzig, JA; Jasiuk, IM; Robling, AG; Turner, CH. Numerical Modeling of Long Bone Adaptation due to Mechanical Loading: Correlation with Experiments. *Ann Biomed Eng*. 2009, 1-11.

[68] Tekkaya, AE; Guneri, A. Shape optimization with the biological growth method: A parameter study. *Engineering Computations* (Swansea, Wales). 1996; 13(8), 4-18.

[69] Heller, M; Kaye, R; Rose, LRF. Gradientless finite element procedure for shape optimization. *J Strain Anal Eng Des*. 1999; 34(5), 323-36.

[70] Das, R; Jones, R; Peng, D. Optimisation of damage tolerant structures using a 3D biological algorithm. *Eng Failure Anal*. 2006; 13, 362-79.

[71] Wu, Z. An efficient approach for shape optimization of components. *International Journal of Mechanical Sciences*. 2005; 47(10), 1595-610.

[72] Viceconti, M; Testi, D; Taddei, F; Martelli, S; Clapworthy, GJ; Van Sint Jan, S. Biomechanics modeling of the musculoskeletal apparatus: Status and key issues. *Proceedings of the IEEE*. 2006; 94(4), 725-38.

[73] Ezquerro, F; Simon, A; Prado, M; Perez, A. Combination of finite element modeling and optimization for the study of lumbar spine biomechanics considering the 3D thorax-pelvis orientation. *Med Eng Phys*. 2004; 26(1), 11-22.

[74] Marom, SA; Linden, MJ. Computer aided stress analysis of long bones utilizing computed tomography. *J Biomech*. 1990; 23(5), 399-404.

[75] Fernandez, JW; Hunter, PJ. An anatomically based patient-specific finite element model of patella articulation: Towards a diagnostic tool. *Biomechanics and Modeling in Mechanobiology*. 2005; 4(1), 20-38.

[76] Phillips, ATM. The femur as a musculo-skeletal construct: A free boundary condition modelling approach. *Med Eng Phys*. 2009; 31(6), 673-80.

[77] Speirs, AD; Heller, MO; Duda, GN; Taylor, WR. Physiologically based boundary conditions in finite element modelling. *J Biomech*. 2007; 40(10), 2318-23.

[78] Fischer, KJ; Jacobs, CR; Carter, DR. Computational method for determination of bone and joint loads using bone density distributions. *J Biomech*. 1995; 28(9), 1127-35.

[79] Be'ery-Lipperman, M; Gefen, A. Contribution of muscular weakness to osteoporosis: Computational and animal models. *Clin Biomech*. 2005; 20(9), 984-97.

[80] Wells, R; Evans, N. Functions and recruitment patterns of one- and two-joint muscles under isometric and walking conditions. *Human Movement Science*. 1987; 6(4), 349-72.

[81] Raikova, RT; Prilutsky, BI. Sensitivity of predicted muscle forces to parameters of the optimization-based human leg model revealed by analytical and numerical analyses. *J Biomech*. 2001; 34(10), 1243-55.

[82] Florio, CS; Narh, KA. Development of a modeling technique for the investigation of muscle activity and its effect on bone stresses in the human leg during an isometric exercise. *Simulation*. 2011; 87(4), 313-33.

[83] Carriero, A; Zavastky, A; Stebbins, J; Theologis, T; Lenaerts, G; Jonkers, I; et al. Stress distributions in the proximal femur during gait in healthy and spastic diplegic children. *55th Annual Meeting of the Orthopedic Research Society*; 2009; Las Vegas, NV.

In: Trabecular and Cortical Bone ISBN: 978-1-62808-766-6
Editor: V. Subramanium, M. Bhatnagar

Chapter V

Advancements in Cortical Bone Imaging

Alexis Donneys[1,*], Noah Nelson[1], Salman Ahsan[1], Jonathan Peterson[1], Michael Morris[2], Francis Esmonde-White[2], Steven R. Buchman[1], Stewart Wang[1] and Benjamin Levi[1]

[1]Department of Surgery, University of Michigan, Ann Arbor, MI
[2]Department of Chemistry, University of Michigan, Ann Arbor, MI

Abstract

Cortical bone is a unique and dynamic tissue due to its complex micro-architecture, structural role in the human skeleton and ability to continuously remodel. Its dense nature makes it clearly distinguishable from other tissues by means of imaging. Through a portrayal of radiodensities, imaging provides both qualitative clues and quantitative metrics reflecting information about cortical integrity, physiology and function. This chapter reviews the imaging of cortical bone from its rudimentary beginnings (X-ray) to current, modern emerging

* Corresponding author: Alexis Donneys, MD, Craniofacial Research Laboratory, 2228, BSRB, 109 Zina Pitcher Place, Ann Arbor, MI 48109-0219; Phone: (215) 219-8748; Fax: (734) 647-0003; alexisd@med.umich.edu.

technologies such as Nano-Computed Tomography and Vascular Microangiography. The clinical importance of cortical bone imaging, as well as its value in research and development is emphasized. Cortical bone morphomics is also highlighted as a potential predictive, rapid clinical screening tool in the modern day trauma bay. The authors further discuss how current innovations in cortical bone imaging and corroborating analytical methodologies such as Raman spectroscopy may lead to useful investigative tools to aid in the early detection of pathologies such as Osteoradionecrosis, Bisphosphonate Related Osteonecrosis of the Jaw (BRONJ) and Heterotopic Ossification.

Background

Much like the walls of a building, cortical bone imparts structural integrity and protection for the inner contents of vital and living tissues. Because of its rigidity, it is often considered inert; however, it is a biologically dynamic tissue capable of methodically responding to the environment. While cortical bone is dense and seemingly inanimate, its dynamic character and responsiveness to stress are often a more important consideration than its immediate structural integrity. Despite our capacity to instantly "see" and evaluate its structural components with modern day imaging, our ability to understand its acute, real-time physiologic processes are still quite constrained. Consider our capability to image the heart or lungs and immediately arrive at conclusions about potential structural and physiologically detrimental pathologic concerns; now juxtapose that with the current inability to discern the same level of understanding regarding ominous bone pathologies. When imaging the heart, all constituents of the image are deemed relevant–the muscle itself, the valves, as well as the vascularity. Clinically, cortical bone imaging has classically focused on its outward structure, with little regard for its inner-workings, such as its complex micro-structural components, vascularity and chemical composition. For this reason, diagnosis and treatment of deleterious cortical bone pathologies largely remain inadequate. The time interval between the onset of a disease process and our ability to visualize its consequences often preclude early intervention and treatment. If our proficiency to image the heart were as rudimentary as that of cortical bone, morbidity, necrosis and death would inevitably ensue. Despite this reality, modern technology at the research bench has afforded us a glimpse of what is possible. For this reason, this chapter reviews innovations in cortical bone imaging, discusses the role of that imaging in the clinic, and also examines the importance of this type of imaging in research and development. Further, we highlight emerging technologies and corroborating analytical methodologies that may lead to useful clinical tools allowing for the early detection and intervention of deleterious bone morbidities that are currently largely undetectable in their initial stages.

Evolution of Cortical Bone Imaging

X-Ray

We can largely credit the beginnings of cortical bone imaging to Wilhelm Conrad Röntgen, a physicist who discovered a powerful use for a particular radiation spectrum while experimenting with cathode ray tubes (Figure 1). We now know that X-rays are a source of radiation with a wavelength spanning between UV and gamma rays (100 to 0.001 nm). However, Röntgen named them X-rays because of an uncertainty of what they were at the time.

Around the turn of the 19th century, physicists became intrigued by the phenomenon surrounding the movement of charged particles through negative pressure environments.

This scientific inquiry led to the discovery of the cathode ray tube. A cathode ray tube is essentially a vacuum tube containing an energy source and a fluorescent screen interposed to collect the images created by the movement of charged particles. The cathode ray tube allows for electromagnetic manipulation of particles prior to acquisition of an image. This technology was instrumental in the discovery of the electron by J. J. Thompson in 1897, an accomplishment that earned him the Nobel Prize in 1906. Paralleling those developments, in 1895, Wilhelm Röntgen came across a distinct peculiarity observed when the cathode tubes were placed in a dark environment completely devoid of light. He found that the rays had a distinct ability to penetrate dense matter, and that the images generated by the penetrating rays could be captured on a fluorescent screen up to a certain distance. He was so intrigued by this phenomenon, that he involved his wife Anne Bertha Ludwig of Zurich in his early experiments. In fact, the first visible evidence of a medical X-ray available today is the image of Bertha's hand demonstrating her highly radiodense bone and wedding ring surrounded by the less dense soft tissue of her hand (Figure 2). For his discovery, Wilhelm Röntgen received the Nobel Prize in 1901 (Nitske, 1971).

Figure 1. A portrait of Wilhelm Conrad Röntgen.

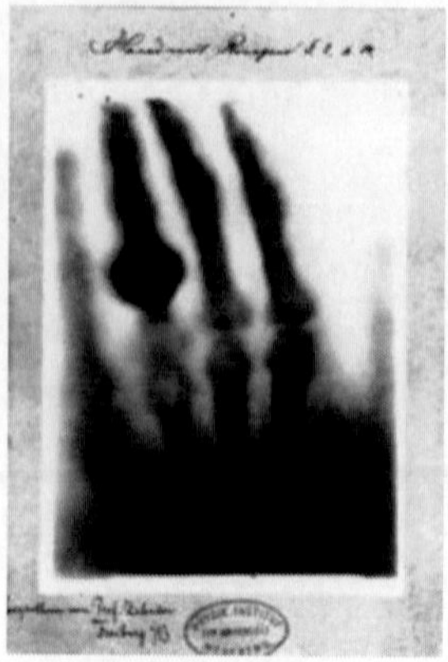

Figure 2. The 1st X-ray taken by Röntgen of his wife Bertha's hand.

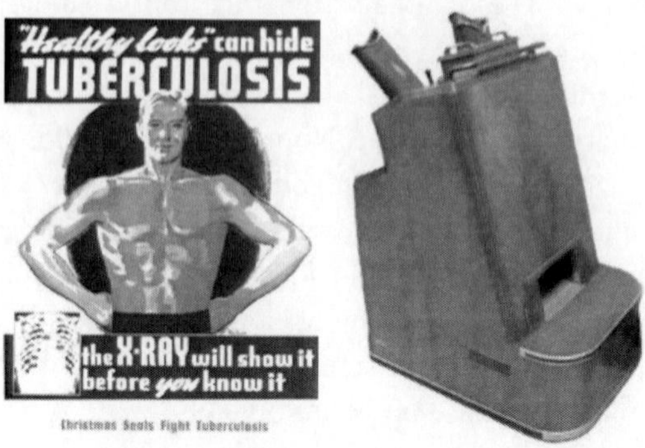

Figure 3. From left to right, public health advertisements about X-rays and detection of tuberculosis, fluoroscope commonly seen in shoe stores of the time.

It soon became customary practice for X-rays to be used widely for Tuberculosis screening. Prior to the discovery of the dangers of ionizing radiation, fluoroscopes were readily available in shoe stores for children to see the bones in their feet as a form of entertainment (Figure 3).

The military also endorsed the use of X-rays, not only as a screening tool, but also as an aid to orthopaedic surgeons to repair fractures and to locate and remove of shrapnel encountered in combat during WWI. In fact, the US Army was instrumental in advancing fluoroscope technology. Fluoroscopy was widely popular since it provided a rapid and efficient means for instantly visualizing internal anatomy without the need for film development (Figure 4). Today the use of X-ray imaging continues to be prevalent, as it is an inexpensive, simple analytical tool that can identify basic problems throughout the body. In fact, seven out of ten people in the U.S. receive health care related X-rays every year accounting for nearly 200 million scans annually (EPA, 2012). X-rays have evolved from large machines occupying entire rooms, to small portable machines that can be brought to the trauma bay, operating room or patient's bedside to provide direct point of care diagnosis. This first line investigative instrument is particularly useful in the field of orthopedics where disruptions in cortical integrity may be rapidly observed and subsequently managed.

Figure 4. Military use of early X-ray devices.

CT, μCT and Present Day Nano-CT

The simultaneous advancement of X-ray innovation and computer development in the first half of the 20th century conferred a repertoire of computational diversity that allowed investigators and inventors to transcend 2D imaging and move to 3D computed tomography (CT). CT is a 3D X-ray imaging technology. Unlike most MRI musculoskeletal applications, CT provides positive contrast of mineralized tissues. The image formation process begins with the acquisition of serial radiographic projections over a range of angular positions around the object of interest. The cross-sectional field of view is then reconstructed using established computational techniques based on Radon projection theory whereby a radon transform represents the scattering data obtained as the output of a tomographic scan. Hence the inverse of the Radon transform can be used to reconstruct the original density from the scattering data, and thus it forms the mathematical underpinning for tomographic reconstruction, also known as image reconstruction. Similar to simple radiography, the reconstructed image intensity values represent the local X-ray attenuation – a material property related to the electron density. Contrast between soft and mineralized tissues in CT is high due to the relative electron-dense inorganic component (calcium hydroxyapatite) of the bone matrix (Krug, 2010).

CT images reconstructed from multiple viewing angles have a significant advantage over conventional radiography images in that they provide quantitative information about the spatial distribution of material density (i.e., linear attenuation coefficient) within an object. Conventional radiography, on the other hand, provides only a summation of material attenuation along the X-ray path. Furthermore, CT completely eliminates the superimposition of images of structures outside the area of interest. Additionally, because of the inherent high-contrast resolution of CT, differences between tissues that differ in physical density by less than 1% can be distinguished. Finally, data from a single CT imaging procedure consisting of either multiple contiguous or one helical scan can be viewed as images in the axial, coronal, or sagittal planes, depending on the diagnostic task. This is referred to as multiplanar reformatted imaging.

While CT scans have important clinical use, their resolution is limited to sizes of about 1.0 mm^3 volume elements. Therefore Microcomputed tomography (μCT), a more sensitive analytic technique, was developed to provide clinicians and researchers with more robust data to study the minute changes in cortical micro-architecture that would be indicators of disease processes. μCT imaging systems have much better spatial resolution, producing voxels that measure 5–50 μm on a side, or approximately 1,000,000 times smaller in volume than CT voxels. Custom built μCT systems were first constructed in the 1980s using microfocal spot X-ray sources and high resolution detectors. These systems were applied primarily as a research tool to quantify structure-function relationships in trabecular bone (Guldberg, 2004).

The range of potential research applications of μCT is both facilitated and limited by methodological and physical factors related to the technology. Some technical requirements can be mutually exclusive, such as increased X-ray source power for dense materials and decreasing micro-focus X-ray source spot size for better spatial resolution. Such challenges necessitate innovative solutions at the forefront of the developing technology. These challenges are addressed by the introduction of the nano-CT scanner. This scanner employs an X-ray source to obtain a spot size of 0.3μm, combined with an extremely high precision sample manipulator and highly sensitive X-ray camera. By permitting high spatial imaging resolution of pixel sizes down to 150 nm, the nano-CT instrument allows clear visualization of structures on the level of cells with appropriate X-ray contrast. Nano-CT allows clear distinction of structures linked to bone quality like osteocytes, microcracks in bone, and resorption lacunae. These sub-micron cracks are of particular importance to biomechanical bone research. The internal ultrastructure of human bone can also be imaged by nano-CT, providing a translational application of this technology (Salmon, 2007). Investigators, to date, have used this promising technology to investigate the differences in morphology of osteocytes and their lacunae in osteoarthritic, osteopenic and osteopetrotic bone (Hove, 2009). In the future this tool has clinically relevant translational potential to better understand the minute micro-structural changes in bone resulting from these disease processes and may potentially permit clinicians to address these changes with targeted pharmacological therapies.

Imaging the Vascularity of Cortical Bone

Vascularity is integral to the structural and physiologic make up of cortical bone, and plays a pivotal role in both normal and pathologic processes. Although the dense nature of cortical bone makes it easily amenable to conventional imaging, visualizing its vascularity provides a considerable challenge. Cortical bone is composed of osteocytes trapped in lacunae, concentrically arranged around a penetrating vessel. These configurations are known as osteons or Haversian systems. Osteocytes communicate via cytoplasmic processes that travel within canaliculi, or interconnecting channels within the bone. These canaliculi receive nutrients from penetrating vessels, and are a means of nutrient transport for the sustenance of the osteocytes. The small size of the Haversian systems and penetrating vessels, as well as the canaliculi within the cortical network and the dense mineral matrix surrounding the vascular channels, produce an obstacle to the successful imaging of the vascularity of cortical bone. The Haversian systems range in size from 100 to 300 μm, while penetrating cortical vessels range in size from 80 to 100 μm and canaliculi are 4 to 10 times smaller ranging in size from 1 to 20 μm (Bone Structure; Pazzaglia, 2008). Despite these challenges, the imaging of cortical vascularity continues to advance both in the clinic and in the research laboratory.

With the use of contrast agents and fluoroscopy, clinicians have a limited potential to image vessels within bone in patients to arrive at important diagnoses regarding injured or compromised bone such as in complex fractures. On the other hand, research investigators have pushed the boundaries of what is possible at the research bench. Rhinelander utilized X-ray microangiography to provide high-resolution 2-D angiograms of dog hind limbs (Rhinelander, 1962). His method of infusing contrast agents and histologically sectioning bone into 1mm thick slices was an instrumental development allowing for the analysis of fracture site vascularity and the relative contributions of vessels from various anatomic areas within the bone to the fracture site (Rhinelander, 1974) (Figure 5 and 6). His investigations offered important contributions in the development of clinical concepts surrounding the topic of medullary nailing for fracture repair (Rhinelander, 1998).

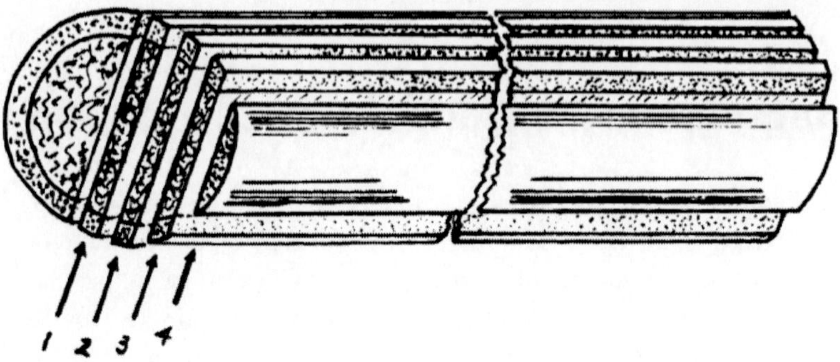

Figure 5. 1mm histologic sections of perfused bone for vascular analysis.

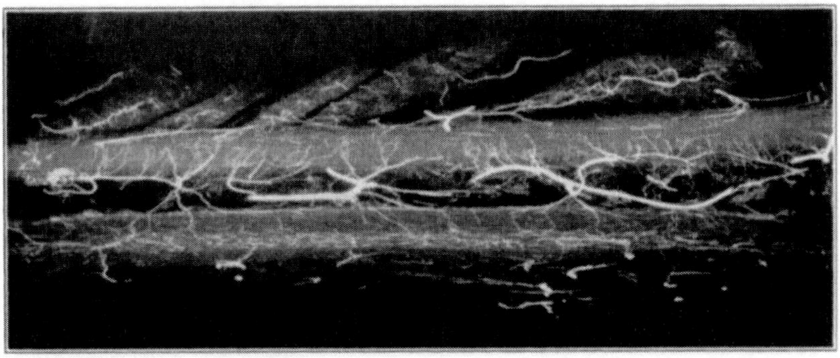

Figure 6. X-ray microangiography of dog hind limbs utilized by Rhinelander from histologic sections.

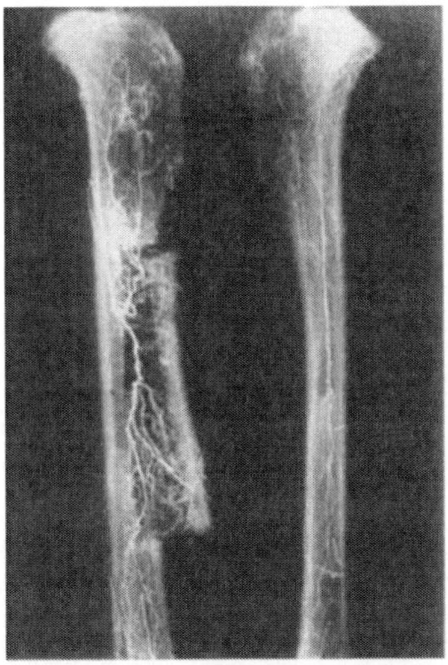

Figure 7. X-ray microangiography of process of transverse distraction osteogenesis utilized by Ilizarov.

Ilizarov also examined the process of Distraction Osteogenesis (DO) with a similar methodology. His important work afforded us a glimpse of the inner workings of the process of distraction. Through his investigations, he discovered that DO was a highly angiogenic process associated with an increased amount of vascularity at the distraction site, beyond that of normal fracture healing. These important concepts are instrumental in the development of novel tissue engineering strategies today (Ilizarov, 1989, 1992) (Figure 7).

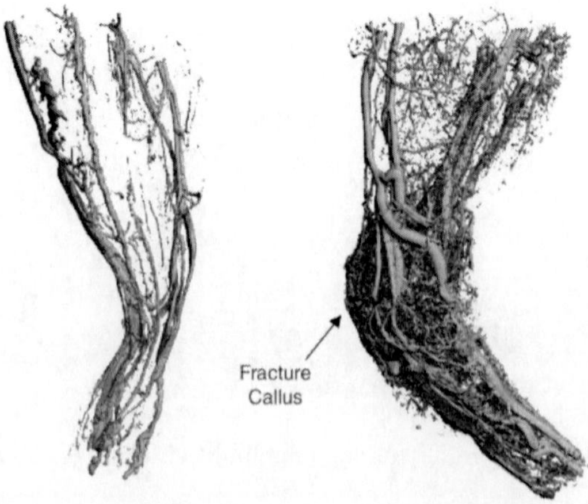

Figure 8. Comparison of vascularity at a fracture site utilizing micro-CT after vascular perfusion.

Although X-ray microangiography provides a high-resolution 2-D view of the vasculature within bone, it lacks the ability to employ a quantitative analysis of vascularity. Guldberg employed a method of vascular perfusion followed by μCT to circumvent these limitations, affording investigators the opportunity to now quantify their observations and arrive at statistically relevant conclusions (Guldberg, 2003). These are powerful analytic tools with particular relevance in the investigation of angiogenic responses following vascular injury, or in emerging tissue-engineering concepts. Guldberg employed this method in animal models designed to examine both long bone fracture healing, and the effect of ischemia on injured tissues (Duvall, 2004) (Figure 8). In this method, either barium sulfate or lead chromate is infused into the vascular tree followed by bone decalcification. This results in a specimen containing a dense outline of the vasculature that is amenable to imaging with μCT. Besides allowing for quantitation, there are unique advantages over 2-D histologic methods, such as the ability to scan at a higher resolution (15-30 μm voxel size), and to employ 3-D connectivity analysis.

Investigators have subsequently employed this useful technique in the examination of bone pathologies with known vascular compromise such as in the development of osteoradionecrosis, pathologic fractures and associated non-unions. Despite these evolutions, 3-D microangiography currently lacks the potential for longitudinal studies due to the need for animal sacrifice upon infusion of the contrast agents (Donneys, 2013) (Figure 9). New contrast agents that allow for in-vivo imaging would be a valuable contribution with relevance to a variety of investigations, such as in the examination of vascularity within implanted tissue engineered constructs over time.

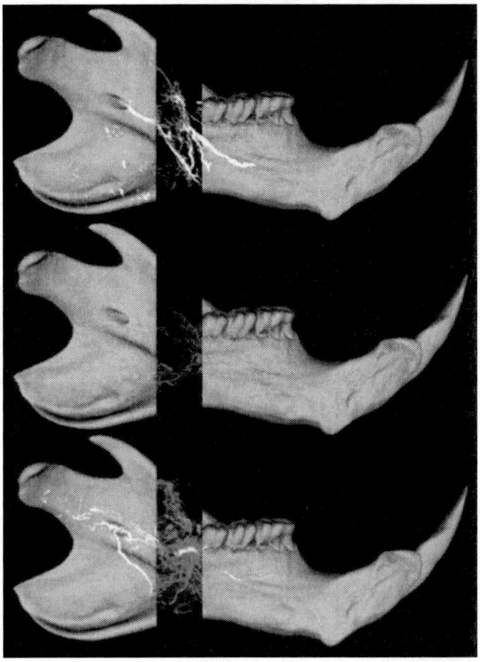

Figure 9. Comparison of vascularity in normal fracture (top), radiated fracture (middle), and radiated fracture treated with angiogenic deferoxamine therapy (bottom) with micro-CT after vascular perfusion.

Traditional and Emerging Radiomorphometrics

The use of μCT for analyzing bone microarchitecture has gained popularity as a rapid, objective method for quantifying data relating to bone density and mineralization. Currently, the popular method for bone analysis involves calculating classic mineralization metrics such as Bone Mineral Density (BMD) and Tissue Mineral Density (TMD) from a scanned region of interest. However, there are inherent known limitations to the exclusive use of these classic radiomorphometrics. For example, while examining osteoporotic risk of fracture, McCreadie reported a large overlap between normal and diseased populations while using these metrics (McCreadie, 2000). While BMD and TMD are indeed indicators of bone mineralization, they are simply reflective of the overall, average mineralization of a bone sample, and they fail to reflect other important quantifiable characteristics of bone biology. In order to address these limitations, Roschger reported using the process of Bone Mineral Density Distribution (BMDD) as a means of examining the entire spectrum of mineralization densities in a sample. He examined a variety of disorders such as osteogenesis imperfecta, osteomalacia, and osteoporosis using human bone biopsy samples in conjunction with high-resolution imaging (Roschger 2008).

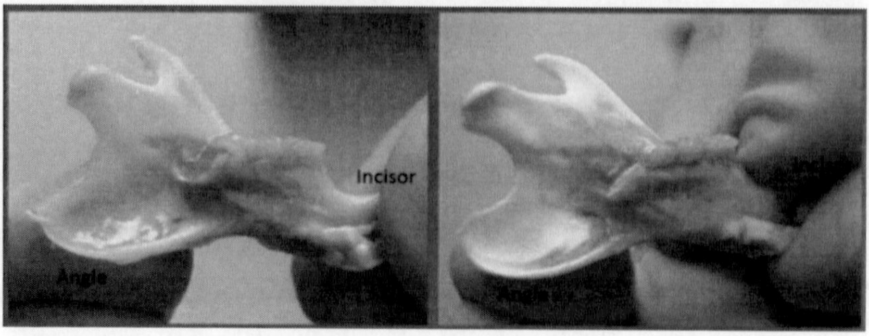

Figure 10. Radiated rat mandible (left) demonstrating cortical thinning in comparison to a non-radiated rat mandible (right).

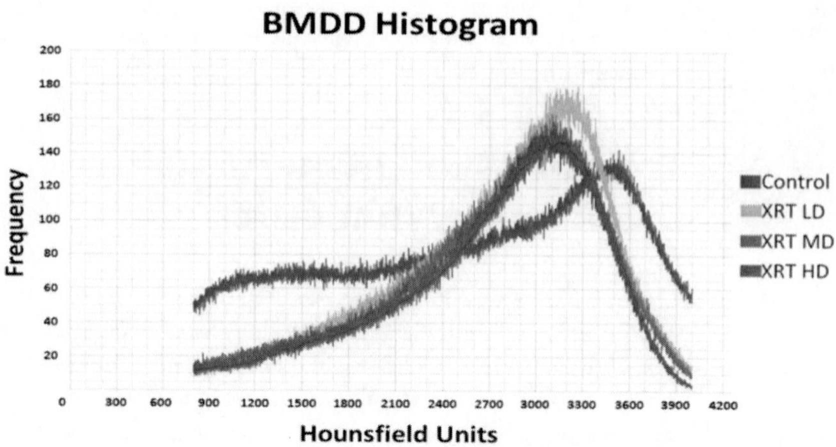

Figure 11. Histogram depicting BMDD for rat mandibles comparing Control (purple), low dose radiation (yellow), medium dose radiation (blue), and high dose radiation (red).

Other investigators experienced similar shortcomings while using the classic radiomorphometrics in the setting of radiotherapy-induced changes to mandibular bone in animal models. When comparing two rat mandible samples (Figure 10) one radiated and one not, the cortical thinning is obvious in the mandibular angle of the radiated sample, to the point where the green glove is visible through the translucent tissue. While it is expected that the TMD value for this sample would be lower, the results were actually the opposite. However, comparison of the BMDD histograms revealed several obvious differences (Figure 11). Shown are the histograms for a control group (purple) and three varying doses of radiation: low (yellow), medium (blue), and high (red). While the TMD was actually higher in radiotherapy than in control, other characteristics such as a notable decrease in bone of low mineralization and a limitation of the range of mineral densities produced were more telling of radiation-induced changes (Donneys, 2012).

At any given point in time within cortico-cancellous bone, there are osteocytes at different stages of the remodeling process, depending on whether they are in the quiescent, activation, or formation phases. This dynamic maintenance results in an overall heterogeneity of mineralization densities throughout a specific volume in a sample of bone. In the context of this non-uniformity, the reliance on metrics such as BMD and TMD that give an average value for a sample can limit the full understanding of the remodeling process. Therefore, applying a more encompassing method of mineral density analysis is a valuable complement. BMDD delivers a wealth of information describing the array of mineralization within a sample. Furthermore, the BMDD metrics may be used to quantify differences in the normative mineralization pattern of different skeletal sites, as well as to note any variability relating to ethnicity, gender, or age [Roschger 2003]. The BMDD histogram can be viewed as a fingerprint representing the complex, dynamic events surrounding bone mineralization at a snapshot in time (Ruffoni, 2007).

A few of the specific metrics that can be ascertained from this histogram are the mineralization peak (M_{peak}), the mineralization width (M_{width}) and the mineralization total number of voxels (M_{tnv}) within a sample. The M_{peak} is a descriptor of bone quality, identifying the most frequently occurring level of mineral density existing in a sample. Changes in the M_{peak} may be indicative of a shifting mineralization density in a skeletal site. The M_{width} is reflective of the diversity of mineralization densities within a selected sample. The M_{width} is directly correlated to the heterogeneity of mineralization of the bone. This metric is especially significant in cases in which bone turnover is limited. In the radiation scenario, bone would become more homogenous throughout a sample, and therefore the M_{width} would decrease as a result. Volumetric changes are comparable with the help of the M_{tnv}. This simple calculation specifies the overall volume of mineral being analyzed. This metric may be used to observe differences in the quantity of mineral between samples. Additionally, this can be used to quickly note changes in mineral volume as a result of exposures or treatment strategies (Donneys, 2012).

Currently, barriers remain to the use of BMDD in a clinical setting. In order to observe the small changes in mineralization that are early indicators of disease, μCT or a similar high resolution imaging modality is superior to the more limited resolution of the clinical CT. Because higher resolution imaging necessitates a higher radiation burden to the patient and longer acquisition times, the wide-spread use of high resolution imaging remains impractical. However, the development of clinical micro-computed tomography (CMCT) is currently underway with targeted utility in superficial skeletal sites and in cochlear implant surgeries involving the intricate anatomy of the inner ear (Wang, 2005). While further clinical optimization will be needed, BMDD may soon be a valuable asset to current diagnostic capabilities.

Morphomics

Recent advances in clinical radiographic modalities and the exciting adaptation of spectroscopic analysis and computational densitometry have greatly improved the accuracy and ability of imaging studies to realize their primary objective: diagnosis. Indeed entire specialties and sub-specialties are dedicated solely to the acquisition and interpretation of the objective data provided by scans to identify abnormalities or rule-out specific diseases. Despite the effectiveness of modern radiologists, however, the power of current imaging modalities yields data-sets in the form of images that contain a wealth of knowledge too vast for even the most proficient observers to fully utilize. CT scans can subdivide an area of the body into hundreds of thin slices allowing for detailed examination of specific anatomy. More broadly, it divides the 3-dimensional body habitus into detailed, smaller pieces that can be analyzed as a whole to comment on the overall health of the individual. Recent, exciting advances in genotyping and the mapping of the entire human genome has provided the opportunity to examine the entire instruction set in an individual instead of searching only for a single, specific abnormality. Similar to this broad genomic approach, discreet cross-sectional images can be analyzed as a whole to yield information about the overall physiologic make-up of an individual, that is, the morphome.

Analytic morphomics makes use of cross-sectional imaging in order to make a more global assessment of patient health, rather than focusing on a specific pathology (Englesbe, 2012). This form of imaging is anatomically indexed based on vertebral bone structures. The process allows for image analysis to move beyond diagnosis and to provide prognostic information. Because imaging studies have become so prevalent in diagnostic medicine, many patients will have extensive radiographic work-ups, especially before elective surgery. Prior to approving surgical procedures, surgeons use several objective factors including age, BMI, and comorbidities to determine if a patient is healthy enough for a procedure. Experienced surgeons also rely on a more subjective assessment of a patient's overall health – the "eyeball test" (Englesbe 2012). Using semi-automated computational indexing of anatomy from cross-sectional images rapidly yields objective data about human morphology. This more objective "eyeball test" can then corroborate or contradict visual assessment of health by providing quantitative data on overall health compared to age.

The process of analytic morphomics began with the automation of computerized indexing of bony landmarks to define anatomic volumes. Then, measurement of cross-sectional area of lean core muscle size was seen to be a more reliable proxy than age or BMI at measuring a patient's risk of surgical complication (Englesbe 2010, Park 2009). This measure of muscle mass for health is commonly used observationally as physicians look for temporal wasting or other signs of sarcopenia as a predictor of patient vigor and physiologic reserve. Several anatomic indices have proven to be better, objective predictors of surgical outcome than commonly used parameters such as age and BMI. These include, for example, the cross-sectional area of the Psoas muscle in patients undergoing treatment for abdominal aortic calcification (Harbaugh, 2013) and abdominal aortic aneurysm repair (Lee, 2011), the size of the temporalis fat pad in patients with craniosynostosis (Rinkinen, 2013), and the thickness of the subcutaneous fat pad in patients undergoing midline laparotomy (Lee, 2011). Other computational analysis of body structures including bone density, muscle-fat ratios, and vascular calcifications are also being studied as markers of frailty, age, and predictors of outcome. Additionally, cortical bone thickness serves as an important predictor of patient age, frailty and risk of injury following trauma. By analyzing large population trends of cortical bone structure, car companies can improve upon vehicle design to decrease stress on specific areas of the skeleton.

Corroborating Analytics and Emerging Innovations

Two major driving factors influence innovations in cortical bone imaging. First, are the clinical limitations for early detection of bone related disease processes, and second, is the quest for optimized and reproducible tissue engineering strategies designed to deliver viable bone tissues from the research bench top to the clinic. The hope is that these tissues will have direct application for filling large bone defects. While cortical bone imaging has largely depended on X-ray or MRI technology, the future landscape will soon increase in complexity to include corroborating emerging technologies such as Near IR and Raman spectroscopy. This complexity is needed because of specific challenges that are unique to bone tissue. Bone has a distinctive remodeling pattern that requires a slow and constant equilibrium of events. Osteoclasts and osteoblasts divide less rapidly than cells responsible for the remodeling and repair of other tissues (Marx, 1987). Therefore, pathologies affecting this cycle are often delayed in presentation, often requiring years to manifest clinically. However, as we begin to consider the chemical composition of bone, and chemical alterations to bone composition due to these pathologies, these emerging technologies may offer a wealth of objective information that is complementary to modern imaging thereby enhancing current diagnostic capabilities.

These future directions are important considerations, as their success would open the door for investigations regarding the potential therapeutic prevention of disorders such as Osteoporosis, Osteoradionecrosis, Bisphosphonate Related Osteonecrosis of the Jaw (BRONJ) and Heterotopic Ossification.

Furthermore, technologies such as these that offer seamless translation from the research laboratory to the clinic would be highly desirable and would increase efficiency for translational developments such as bone engineered constructs. For these reasons, we have included a review of emerging corroborating technologies in this chapter.

Near IR and Raman Spectroscopy

Raman and infrared spectroscopy measure molecular vibrations. These are bending and stretching motions of the bonds that connect the atoms of a molecule, or of an ion containing at least two atoms. These vibrations occur at specific frequencies. They can be observed as absorption of light in specific regions of the infrared spectrum or as shifts in the wavelength of scattered laser light. Infrared absorption is called Fourier transform infrared (FTIR) spectroscopy. The light scattering technique is called Raman spectroscopy (Paschalis, 2011; Morris, 2011).

Historically, correlations between spectral features and bone composition and physical properties were developed first in infrared spectroscopy, mostly in the 1980's and early 1990's. Raman spectroscopy was used in bone studies only sporadically until the early 2000's. The interpretation of Raman spectra regarding bone matrix collagen depends heavily on correlations already developed for infrared spectra, because the spectra have many bands in common. Theory predicts that for oxy-anions (phosphate, carbonate), bands that are strong in the infrared will be weak in the Raman and vice versa. Therefore interpretation of bone mineral bands is derived from extensive studies by chemists and geologists of phosphate and carbonate minerals and synthetic calcium phosphates and calcium carbonates (Table 1).

Both techniques provide information on composition of bone. Importantly, the exact band position and width depends on more than just composition. In bone mineral, the bandwidths are measures of crystallinity, a parameter that describes crystallite size and perfection (increasing heterogeneity makes the bands wider). The presence of a large amount of monohydrogen phosphate in newly modeled bone mineral also shifts the main phosphate band position and broadens it in comparison to the phosphate band in mature mineral. Similarly, collagen secondary structure is perturbed by the presence of cross-links. As bone matures the fraction of non-reducible cross-links increases and the ratio of two components of the amide carbonyl band (at about 1650 cm^{-1}) change intensity, as a result of changing collagen secondary structure. Measures of cross-linking are indirect and so are open to multiple, context-dependent interpretations. For example, it has been shown by Raman spectroscopy, that amide I component intensities at fracture boundaries change in the direction associated with reduced cross-linking. Moreover, instrumentation is compact and portable and uses low intensity near-infrared radiation. Raman employs fiber optic probes that can be hand-held and can be stand-off, allowing for direct observation of affected sites that might otherwise be sensitive to pressure or physical manipulation during the exam. This technique, in fact, will provide an ideal screening tool, as it is non-ionizing, non-contact, and has high sensitivity for the types of mineralization events that are characteristic of heterotopic ossification.

Table 1. Raman band assignments and positions for bone, from literature sources

Band Assignment	Position, cm^{-1}
Phosphate P-O stretch ν_1	957-962
Mineral crystallinity	1/FWHM ν_1
Carbonate C-O ν_1	1071
Hydroxyproline ring	876
Proline ring	916
Phenylalanine, ring breathing	1003
Amide III	1243-1269
CH_2 deformation	1447-1452
Amide I	1660-1690
Cross-links ratio	1660/1690

Near-infrared (NIR) optical imaging can visualize multiple contrast agents at nanomolar levels due to high tissue penetration and low tissue autofluorescence at this spectral region. Several laboratories have described the use of NIR labeled bisphosphonates, tetracyclines, and activatable substrates as markers for the primary mediators of bone metabolism. These contrast agents have been used to directly visualize early mineralization associated with atherosclerotic plaques prior to identification by more traditional markers. The administration of tetracyclines to denote regions of local bone mineralization was first described in 1958 by Milch and co-authors (Milch 1958) and subsequently popularized by Frost for use in quantifying bone turnover indices from tissue biopsies (Frost, 1961). However, tetracycline use was dominantly restricted to histologic sections due to limitations in light penetration and tissue autofluorescence in the spectral regions most associated with tetracycline use. The advent and use of near-infrared fluorophores has removed many of these restrictions, allowing for non-invasive *in vivo* imaging of contrast agents at very low concentrations. While radiography is able to image individual contrast agents at millimolar concentrations, near-infrared optical imaging can visualize multiple contrast agents at nanomolar levels (Pogue 2010) due to high tissue penetration of light and low tissue autofluorescence within this region of the spectrum (Weissleder 2003; Kozloff 2009; Kozloff 2010; Kozloff 2007). These contrast agents have been used to directly visualize early mineralization associated with atherosclerotic plaques prior to identification by more traditional markers (Aikawa 2007), and loss of bone mineralization in models of atherosclerosis and chronic renal disease (Hjortnaes 2010). Similar near-infrared, protease-sensitive, activatable contrast agents have been used to detect MMP9 expression as an early marker for heterotopic ossification (Azhdarinia 2011; Rodenberg 2011), but these techniques have not yet been applied to the mineralization events themselves.

FTIR has found greatest use in studies of bone quality. In an early application of FTIR imaging, it was shown that osteoporotic cancellous bone contained larger crystallites than healthy bone from matched controls, and that the distribution of crystallite sizes was narrower in the osteoporotic tissue. This information is not available from X-ray based techniques such as computed tomography. More recently, FTIR spectroscopy has shown that in insulin-like growth factor I (IGF-I) deficient mice, mineralization of the development is inhibited, as was known from conventional imaging, but mineral crystallinity is also lower than in the wild type mouse. The effects of IGF-I, therefore, are subtler than just promotion of mineralization.

Raman spectroscopy has been used for similar studies. Recently, the technique has been used to characterize radiation damage to bone caused by anti-cancer therapies, and to demonstrate protection of the tissue with Amifostine. Similarly, Raman studies have also characterized heterotopic ossification signatures, which at early stages are characteristic of newly modeled bone. Raman spectroscopy has potential for a wide range of applications because fresh, hydrated tissue can be used and *in vivo* subsurface spectroscopy and low definition mapping are feasible. At this writing, *in vivo* bone Raman spectroscopy using fiber optic Raman probes has been demonstrated in anaesthetized rodents (Figure 12).

Validation studies show that through-the skin and exposed bone spectra agree quantitatively within about 1%. Similarly, preliminary reports of through-the-skin Raman spectroscopy have appeared. Low definition tomography has been demonstrated on animal limbs. There are already research conference reports describing validation studies on human subjects, in which pre-operative non-invasive measurements are made prior to orthopaedic procedures such as ACL reconstruction and validated by spectroscopy of exposed bone during the procedure.

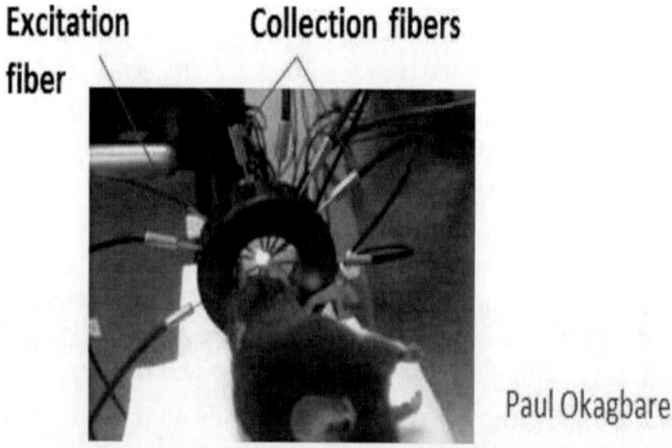

Figure 12. Raman Halo developed to allow for transcutaneous analysis of heterotopic ossification placed on a mouse hind limb.

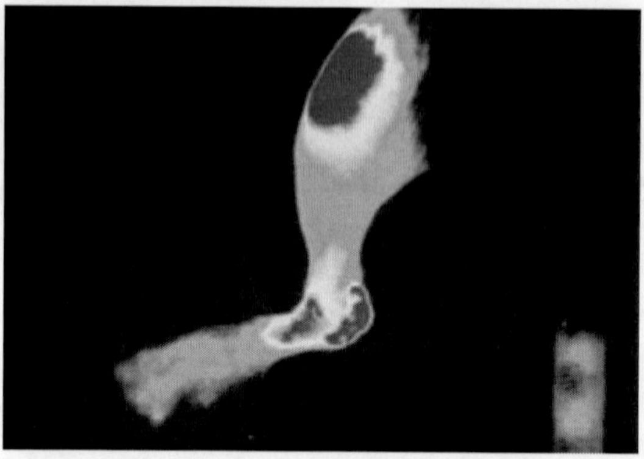

Figure 13. Tetracycline labeled Near IR demonstrating early bone deposition after an Achilles tenotomy and burn injury in a murine model.

In summary, because they provide valuable information on the composition and ultrastructure of excised bone specimens, Raman and infrared spectroscopies are already important tools in basic bone research, and can be viewed as adjuncts to conventional imaging and histopathology. The next few years will likely be a period of intense translational activity for these methodologies. Specialists are developing Raman spectroscopy and FTIR instruments for diagnosis of bone pathologies from tissue specimens, and the development of fiber optic Raman probes for non-invasive diagnostics at the bedside is underway (Figure 13).

In conclusion, over the past century advancements in cortical bone imaging have allowed researchers and physicians alike to further delineate the micro-anatomy and physiology of cortical bone. While the X-rays of the early 20th century allowed for visualization of the cortical surface, today, with modalities such as nano-CT we are able to distinguish structures linked to bone quality such as osteocytes, microcracks and resorption lacunae. Metrics of bone quality such as Bone Mineral Density Distribution have become more encompassing. Angiography has also advanced to allow for visualization of the microvasculature within bone. Further, corroborating analytics such as Raman spectroscopy and Near IR, in conjunction with modern imaging have made it possible to investigate the chemical composition of cortical bone in normal and diseased states. These advancements taken together may make it possible in the near future to detect pathologies such as Osteoradionecrosis, Bisphosphonate Related Osteonecrosis of the Jaw (BRONJ) and Heterotopic Ossification in their early stages, thus enabling the potential for preventative intervention.

References

"Bone Structure." (2013). *Bone Structure*. University of Michigan. Accessed 28 May.

"Medical X-Rays." (15 Nov. 2012). *EPA.gov*. Environmental Protection Agency, Accessed 28 May 2013.

Aikawa, E. et al. (2007). Osteogenesis associates with inflammation in early-stage atherosclerosis evaluated by molecular imaging in vivo. *Circulation, 116*(24), 2841-2850.

Azhdarinia, A. et al. (2011). Characterization of chemical, radiochemical and optical properties of a dual-labeled MMP-9 targeting peptide. (Translated from eng) *Bioorg Med Chem, 19*(12), 3769-3776 (in eng).

Donneys, A., Nelson, N. S., Deshpande, S. S., Boguslawski, M. J., Tchanque-Fossuo, C. N., Farberg, A. S. & Buchman, S. R. (2012). Quantifying Mineralization Using Bone Mineral Density Distribution in the Mandible. *Journal of Craniofacial Surgery, 23*(5), 1502-1506.

Donneys, A., Weiss, D. M., Deshpande, S. S., Ahsan, S., Tchanque-Fossuo, C. N., Sarhaddi, D., Levi, B., Goldstein, S. A. & Buchman, S. R. (2013). Localized deferoxamine injection augments vascularity and improves bony union in pathologic fracture healing after radiotherapy. *Bone, 52*(1), 318-325.

Duvall, C. L., Taylor, W. R., Weiss, D. & Guldberg, R. E. (2004). Quantitative microcomputed tomography analysis of collateral vessel development after ischemic injury. *American Journal of Physiology-Heart and Circulatory Physiology*, *287*(1), H302-H310.

Englesbe, M. J. (2012). Quantifying the eyeball test: Sarcopenia, analytic morphomics, and liver transplantation. *Liver Transplantation*, *18*(10), 1136-1137.

Englesbe, M. J., Lee, J. S., He, K., Fan, L., Schaubel, D. E., Sheetz, K. H. & Wang, S. C. (2012). Analytic morphomics, core muscle size, and surgical outcomes. *Annals of surgery*, *256*(2), 255-261.

Englesbe, M. J., Patel, S. P., He, K., Lynch, R. J., Schaubel, D. E., Harbaugh, C. & Sonnenday, C. J. (2010). Sarcopenia and mortality after liver transplantation. *Journal of the American College of Surgeons*, *211*(2), 271-278.

Frost, H. M., Villanueva, A. R., Roth, H. & Stanisavljevic, S. (1961). Tetracycline bone labeling. (Translated from eng) *J New Drugs*, *1*, 206-216 (in eng).

Guldberg, R. E., Ballock, R. T., Boyan, B. D., Duvall, C. L., Lin, A. S. P., Nagaraja, S. & Taylor, W. R. (2003). Analyzing bone, blood vessels, and biomaterials with microcomputed tomography. *Engineering in Medicine and Biology Magazine*, IEEE, *22*(5), 77-83.

Guldberg, R. E., Lin, A. S., Coleman, R., Robertson, G. & Duvall, C. (2004). *Microcomputed tomography imaging of skeletal development and growth.* Birth Defects Research Part C: Embryo Today: Reviews, *72*(3), 250-259.

Harbaugh, C. M., Terjimanian, M. N., Lee, J. S., Alawieh, A. Z., Kowalsky, D. B., Tishberg, L. M. & Englesbe, M. J. (2013). Abdominal Aortic Calcification and Surgical Outcomes in Patients With No Known Cardiovascular Risk Factors. *Annals of Surgery*, *257*(4), 774-781.

Hjortnaes, J. et al. (2010). Arterial and aortic valve calcification inversely correlates with osteoporotic bone remodelling: a role for inflammation. (Translated from eng) *Eur Heart J*, *31*(16), 1975-1984 (in eng).

Ilizarov, G. A. & Ledyaev, V. I. (1992). The replacement of long tubular bone defects by lengthening distraction osteotomy of one of the fragments. *Clinical orthopaedics and related research*, *280*, 7-10.

Ilizarov, G. A. (1989). The tension-stress effect on the genesis and growth of tissues: Part I. The influence of stability of fixation and soft-tissue preservation. *Clinical orthopaedics and related research*, *238*, 249-281.

Kozloff, K. M. et al. (2009). Non-invasive optical detection of cathepsin K-mediated fluorescence reveals osteoclast activity in vitro and in vivo. *Bone*, *44*(2), 190-198.

Kozloff, K. M., Volakis, L. I., Marini, J. C. & Caird, M. S. (2010). Near-infrared fluorescent probe traces bisphosphonate delivery and retention in vivo. (Translated from eng) *J Bone Miner Res*, *25*(8), 1748-1758 (in eng).

Kozloff, K. M., Weissleder, R. & Mahmood, U. (2007). Non-invasive optical detection of bone mineral. *J Bone Miner Res*, *22*(8), 1208-1216.

Krug, R., Burghardt, A. J., Majumdar, S. & Link, T. M. (2010). High-resolution imaging techniques for the assessment of osteoporosis. *Radiologic clinics of North America*, *48*(3), 601.

Lee, J. S., Harbaugh, C. M., Terjimanian, M. N., Tishberg, L. M., Alawieh, A. Z., Kowalsky, D. B. & Englesbe, M. J. (2011). Abdominal Aorta Calcification In Surgical Patients. *Journal of Surgical Research*, *165*(2), 332.

Lizarov, G. A. (1989). The tension-stress effect on the genesis and growth of tissues: Part II. The influence of the rate and frequency of distraction. *Clinical orthopaedics and related research*, *239*, 263-285.

Marx, R. E. & Johnson, R. P. (1987). Studies in the radiobiology of osteoradionecrosis and their clinical significance. Oral surgery, Oral medicine, *Oral pathology*, *64*(4), 379-390.

Mccreadie, B. R. & Goldstein, S. A. (2000). Biomechanics of fracture: is bone mineral density sufficient to assess risk?. *Journal of Bone and Mineral Research*, *15*(12), 2305-2308.

Milch, R. A, Rall, D. P. & Tobie, J. E. (1958). Fluorescence of tetracycline antibiotics in bone. (Translated from eng) *J Bone Joint Surg Am*, *40*-A(4), 897-910 (in eng).

Morris, M. D. & Mandair, G. S. (2011). Raman assessment of bone quality. *Clinical orthopaedics and related research*, *469*(8), 2160.

Nitske, Robert W. (1971). *The Life of W. C. Röntgen, Discoverer of the X-Ray*, University of Arizona Press.

Park, S. L., Goodman, M. T., Zhang, Z. F., Kolonel, L. N., Henderson, B. E., & Setiawan, V. W. (2010). Body size, adult BMI gain and endometrial cancer risk: the multiethnic cohort. *International Journal of Cancer*, *126*(2), 490-499????

Paschalis, E. P., Mendelsohn, R. & Boskey, A. L. (2011). Infrared assessment of bone quality: a review. *Clinical orthopaedics and related research*, *469*(8), 2170.

Pazzaglia, U. E., Bonaspetti, G., Ranchetti, F. & Bettinsoli, P. (2008). A model of the intracortical vascular system of long bones and of its organization: an experimental study in rabbit femur and tibia. *Journal of Anatomy*, *213*(2), 183-193.

Pogue, B. W., Leblond, F., Krishnaswamy, V. & Paulsen, K. D. (2010). Radiologic and near-infrared/optical spectroscopic imaging: where is the synergy? (Translated from eng) *AJR Am J Roentgenol*, *195*(2), 321-332 (in eng).

Rhinelander, F. W. & Baragry, R. A. (1962). Microangiography in bone healing I. Undisplaced closed fractures. *The Journal of Bone & Joint Surgery*, *44*(7), 1273-1298.

Rhinelander, F. W. & Peltier, L. F. (1998). Effects of medullary nailing on the normal blood supply of diaphyseal cortex. *Clinical orthopaedics and related research*, *350*, 5-17.

Rhinelander, F. W. (1974). Tibial blood supply in relation to fracture healing. *Clinical orthopaedics and related research*, *105*, 34-81.

Rinkinen, J., Zhang, P., Wang, L., Enchakalody, B., Terjimanian, M., Holcomb, S. & Levi, B. (2013). Novel Temporalis Muscle and Fat Pad Morphomic Analyses Aids Preoperative Risk Evaluation and Outcome Assessment in Nonsyndromic Craniosynostosis. *Journal of Craniofacial Surgery*, *24*(1), 250-255.

Rodenberg, E. et al. (2011). Matrix metalloproteinase-9 is a diagnostic marker of heterotopic ossification in a murine model. (Translated from eng) *Tissue Eng Part A*, *17*(19-20), 2487-2496 (in eng).

Roschger, P., Gupta, H. S., Berzlanovich, A., Ittner, G., Dempster, D. W., Fratzl, P. & Klaushofer, K. (2003). Constant mineralization density distribution in cancellous human bone. *Bone*, *32*(3), 316-323.

Roschger, P., Paschalis, E. P., Fratzl, P. & Klaushofer, K. (2008). Bone mineralization density distribution in health and disease. *Bone*, *42*(3), 456-466.

Ruffoni, D., Fratzl, P., Roschger, P., Klaushofer, K. & Weinkamer, R. (2007). The bone mineralization density distribution as a fingerprint of the mineralization process. *Bone*, *40*(5), 1308-1319.

Salmon, P. L. & Sasov, A. Y. (2007). Application of nano-CT and high-resolution micro-CT to study bone quality and ultrastructure, scaffold biomaterials and vascular networks. In Advanced Bioimaging Technologies in Assessment of the Quality of Bone and Scaffold Materials (323-331). Springer Berlin Heidelberg.

van Hove, R. P., Nolte, P. A., Vatsa, A., Semeins, C. M., Salmon, P. L., Smit, T. H. & Klein-Nulend, J. (2009). Osteocyte morphology in human tibiae of different bone pathologies with different bone mineral density—Is there a role for mechanosensing?. *Bone*, *45*(2), 321-329.

Wang, G., Zhao, S., Yu, H., Miller, C. A., Abbas, P. J., Gantz, B. J. & Rubinstein, J. T. (2005). Design, analysis and simulation for development of the first clinical micro-CT scanner. *Academic radiology*, *12*(4), 511-525.

Weissleder, R. & Ntziachristos, V. (2003). Shedding light onto live molecular targets. *Nat Med*, *9*(1), 123-128.

In: Trabecular and Cortical Bone ISBN: 978-1-62808-766-6
Editor: V. Subramanium, M. Bhatnagar © 2014 Nova Science Publishers, Inc.

Chapter VI

Morphology of the Pelvis (Acetabulum) and Femur

***Masaaki Maruyama*[1,*], *Judy R. Feinberg*[2,†], *William N. Capello*[2] and *James A. D'Antonio*[3]**
[1]Department of Orthopaedic Surgery, Shinonoi General Hospital, Shinonoi, Nagano City, Japan
[2]Department of Orthopaedic Surgery, Indiana University Medical Center, Indianapolis, IN, US
[3]GPOA, Sewickley, PA, US

Abstract

Introduction: There are gender differences between male and female in morphology of the pelvis and femur. To confirm this finding, we studied the pelves and femurs in Department of Anthropology, Cleveland Natural History of Museum.

* Masaaki Maruyama, M.D., Ph.D. Department of Orthopaedic Surgery, Shinonoi General Hospital, 666-1 Ai, Shinonoi, Nagano City, Nagano Prefecture 388-8004, Japan. Tel: (81+26) 292-2261; Fax: (81+26) 293-0025.

† Judy R. Feinberg, PhD, Retired Faculty, and William N. Capello, MD. Department of Orthopaedic Surgery, Indiana University Medical Center, Clinical Building Suite 600, 541 Clinical Drive, Indianapolis, IN 46202-5111, US. Tel: (317) 274-8617; Fax: (317) 274-3702.

Materials and Methods: The bones of 50 males and 50 females (total 200 hip joints) were selected from 3,300 skeletons in the museum at random under the following condition: the same distribution of age between the two groups, such as 5 (under 40), 5 (40-49), 15 (50-59), 15 (60-69), 10 (70 or more); and 20% of black in each age class of the groups. The notch acetabular angle was defined as an angle between a line connected the sciatic notch with the posterior acetabular ridge and a line connected both the posterior and anterior acetabular ridge. Each anteversion angle and inclination of the acetabulum, and antetorsion angle of the femur was also measured and statistically analyzed. The results were analyzed by unpaired *t*-test and considered *p* values of less than 0.05 to be significant. Femoral configuration, such as antetorsion of the femoral neck, neck shaft angle, medial offset of the femoral head, anterior bowing and lateral bowing of the femoral shaft was also examined in direct measurement and in radiographs.

Results: The shape of the anterior column of the acetabulum was classified in four types, such as straight, curved, angular, and irregular. Most of the pelves have the curved configuration (60.5%). The notch-acetabular angle was 88.4 ± 3.0° in male and 89.6 ± 3.8° in female ($p=0.013$). All of the angles were ranged within 90 ± 10°. The anteversion angle of the acetabulum was 18.5 ± 5.8° in male and 21.3 ± 7.1° in female ($p=0.002$). The antetorsion angle of the femoral neck was 9.8 ± 9.0° in male and 9.8 ± 8.0° in female ($p=0.954$). There were no correlations between the notch acetabular angle and the other angles ($r^2<0.08$). The neck shaft angle of the femur in neutral rotation was 123.2 ± 6.3° in male and 125.3 ± 5.4° in female ($p=0.014$). The neck shaft angle of the femur in derotation was 121.8 ± 6.1° in male and 123.8 ± 5.2° in female ($p=0.0158$). The medial offset of the femoral head in neutral rotation was 47.2 ± 6.4 mm in male and 41.9 ± 5.9 mm in female ($p<0.001$). The medial offset of the femoral head in derotation was 50.1 ± 5.3 mm in male and 44.3 ± 5.4 mm in female ($p<0.001$). The anterior bowing and lateral bowing of the femoral shaft was 10.3 ± 2.0° in male and 9.0 ± 2.4° in female ($p<0.001$).). The lateral bowing of the femoral shaft was 15.7 ± 8.7° in male and 10.1 ± 9.7° in female ($p<0.001$). There were three distinct femoral shaft configurations, such as lateral, double, and medial curve, of femoral shaft in neutral rotation, in anterior-posterior view.

Discussion and Conclusion: Dimensional differences of the pelvis and femur between females and males were significant with a wide variation within sexes. Most of acetabula had a non-straight acetabular ridge. Configuration of the ridge cannot be detected on standard radiographs nor intraoperatively. Measurement of anteversion angle is dependent upon the point of measurement along the acetabular ridge. Anteversion angle could easily be over-estimated in cases of non-straight anterior acetabular ridge. The notch acetabular angle had small standard deviations, such as 3.0° or 3.8°, and an extremely small range of 90 ± 10° in all of the cases.

Although female has slightly greater angle than male, the difference was only 1.2°. For the cases with posterior osteophyte of the acetabulum, this angle was decreased but more than 80°. In addition, this angle was not correlated with the other parameters. As a result, the notch acetabular angle was useful to identify the anteversion of acetabulum intraoperatively. Medial offset measurement assumes straight long axis of femur. Standard radiographs of hip include only proximal femur from which the long axis cannot be accurately determined. Offset measurement was influenced by rotation of femur. The actual amount of offset may be under-estimated on pre-operative radiographic assessment. Effect of lateral bowing on the apparent long axis of femur was a change of +3.5° varus. Similarly, bowing of the femur was found to be more complex than previously thought. In addition to anterior bowing, some femurs were found to have lateral and even medial bowing. Dimensional differences of the pelvis and femur between females and males were significant with a wide variation within sexes.

Introduction

It is useful to know the normal configurations of acetabulum and femur for clinical application, especially acetabular and femoral anteversion for proper implant positioning and to prevent dislocation in total hip arthroplasty. [2, 10] Adequate anteversion of the hip prostheses provides for a satisfactory functional range of motion. The point of measurement along a possibly curved or angular configuration seems to influence on the acetabular anteversion angle, since the acetabulum is not a simple hemispheric shape. Neither standard radiographs nor two-dimensional computed tomography (CT) scans allow for accurate determination of anatomic configuration [3] and, therefore, assuming acetabular wall configuration has some influence, accurate measurement of the anatomic anteversion angle cannot be determined. Three-dimensional CT scanning has potential for detection of variations in configuration, [1] although this technology currently is not used routinely nor is it cost-effective for routine use as a preoperative assessment tool to aid in intraoperative determination of anteversion. Despite the potential importance of acetabular ridge features, there have been no reports on how differences in anatomic structure influence measurement of acetabular anteversion.

A similar problem exists for accurate measurement of the medial offset of the femoral head. Offset measurement is greatly affected by femoral anteversion and bowing, neither of which can be appreciated adequately on two dimensional radiographs.

Accurate assessment is complicated by the fact that most patients with advanced osteoarthritis of the hip cannot internally rotate their hip on radiographs obtained preoperatively and therefore, the offset is influenced by the inability to attain the necessary derotation of the femur. Several studies of proximal femoral anatomy have regarded the femoral shaft as straight in the standard anteroposterior (AP) view. [8, 22, 25] It is well known that the femoral shaft has some anterior bowing and clinically other configurations have been seen.

The overall incidence of dislocation after primary total hip arthroplasty averages between 2% and 3%. [19] Several risk factors have been identified. There is a definite predilection in female patients with some studies reporting rates double that of male patients. [10, 12, 20, 29] Component positioning has been linked directly to incidence of dislocation. Fackler and Poss [6] reported a significant increase in postoperative dislocations when femoral offset was decreased by inserting the femoral component with a valgus neck shaft angle. Cup orientation is particularly critical. Lewinnek et al. [13] described a safe zone of acetabular component orientation as a lateral opening of 40°± 10° with anteversion of 15°± 10° to decrease the likelihood of dislocation after total hip arthroplasty. Surgical approach also can affect proper positioning of the acetabular component. For example, it has been shown that the surgeon tends to place the cup in 5° to 7° less anteversion with a posterior approach. [18] Positioning the patient in the lateral decubitus position during total hip arthroplasty can allow the pelvis to rotate forward and result in placement of the cup in a retroverted position. [19] Yet, despite all of these reports as to the importance of controlling the amount of anteversion, there have been no reports of how to measure anteversion in an accurate and consistent manner. External alignment guides have been shown to be helpful in reproducing abduction but inaccurate regarding anteversion of the acetabular component. [7] Recently a computer- assisted and image-guided system (HipNav, Center for Orthopaedic Research, Pittsburgh, PA) was developed and is being tested clinically for precise planning and placement of the acetabular component in total hip replacement surgery. [9] In total hip arthroplasty, it was extremely difficult to install the socket in coinciding with anteversion of acetabulum. Dr. D'Antonio suggested that the notch-acetabular angle was nearly equal 90. To confirm this finding, we studied the pelves and femurs in Department of Anthropology, Cleveland Natural History of Museum.

Items of measurements and evaluations in the study are listed in the table 1. The aims of the current study were to: (1) evaluate in detail the morphologic features of the pelvis and femur particularly for configuration of the acetabular anterior and posterior columns and the medial offset of the femoral head relative to the configuration of the femoral shaft; (2) evaluate any differences attributable to gender; and (3) determine whether there is any practical way to accurately assess acetabular anteversion intraoperatively either directly or indirectly.

Materials and Methods

This study was done in collaboration with the Department of Physical Anthropology at the Cleveland Museum of Natural History. The department houses the Hamann-Todd Osteological Collection, which was amassed between 1912 and 1938 by Drs. CA. Hamann and T.W. Todd of the Western Reserve University, Department of Anatomy, and it contains 3100 skeletons of the unclaimed dead of the Cleveland area. [17] Only normal pelves and femurs were included for this study. Specimens with osteoarthritis of the hip, evidence of previous trauma to the pelvis or femur, or skeletal disorders such as multiple epiphyseal dysplasia or osteogenesis imperfecta were excluded. Because gender differences were considered a primary objective of the study, the sample group consisted of 50 males and 50 females (total 200 hip joints) skeletons were selected from 3,300 skeletons in the museum.

The age and race distribution selected to be similar to the population having total hip arthroplasty in the United States. Specimens were selected randomly based on the following conditions: five were younger than 40 years, five were between 40 and 49 years, 15 were between 50 and 59 years, 15 were between 60 and 69 years, and 10 were 70 years or older. Twenty percent of the skeletons in each age group were from African-Americans (black). The average age was 57.9 ± 12.2 years (range, 28-82 years) for males and 57.5 ± 13.5 years (range, 18-82 years) for females. Two hundred normal femur and pelvis combinations (100 male, 100 female) were studied. All measurements were done by one author (MM). [14]

The individual bones (iliums, sacrums, and femurs) were stored in separate areas in the museum, and therefore required reconstruction for this study. The pelves were reconstructed using a thin clay (0.5 to 1.5 cm, proportional to the size of the pelvis) [27] and rubber bands. An anatomic frontal plane of the pelvis was defined by the anterosuperior iliac spines and by the pubic symphysis. This plane is nearly vertical during upright sitting, standing, and walking. [1, 16] In this study, the pelvis was laid on a flat glass table (craniograph) in a prone position so that the anterosuperior iliac spines and the pubic symphysis were in contact with the glass. The glass table was regarded as the anatomic frontal plane (Figure 1).

Table 1. Items for Pelvic and Femoral Measurements

<Pelvis (Acetabulum)>

- Acetabular ridge configuration
- Acetabular anteversion angle
- Notch acetabular angle

<Femur>

- Femoral shaft configuration
- Femoral anterior and lateral bowing angles
- Femoral anteversion angle
- Femoral neck shaft angle and medial offset

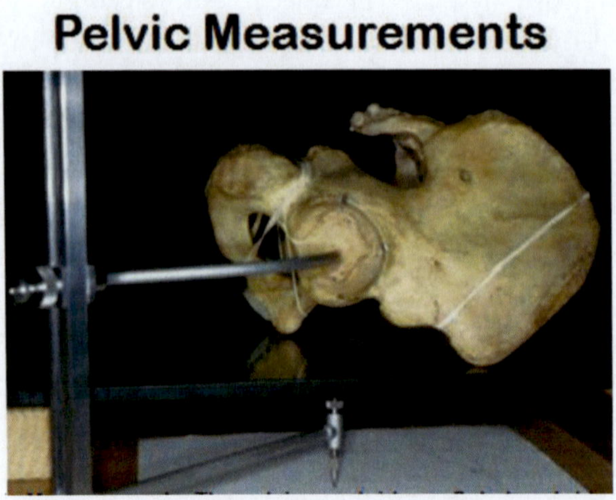

Figure 1. Pelvic Measurements. The pelvis was laid on a flat glass table (craniograph) in a prone position so that the anterosuperior iliac spines and the pubic symphysis were in contact with the glass. The glass table was regarded as the anatomic frontal plane.

The femur was placed on an osteometric board in the supine position so that the posterior aspects of the medial and lateral femoral condyles and the base of the femoral neck were in direct contact with the board (Figure 2).

Statistical Analysis

Statistical analyses for comparison of morphologic features between males and females were done using unpaired *t*-tests or chi square analysis. Comparisons from right to left femur were done using paired *t*-tests. Correlations between variables were calculated using the Pearson product moment coefficient of correlation (r).

The results were analyzed and considered p values of less than 0.05 to be significant. The statistical package used was Microsoft® Excel and its statistical software (Microsoft Corporation, Redmond, WA).

Pelvis (Acetabulum)

Morphologic Features

The shape of the anterior column of the acetabulum was classified in four types, such as straight, curved, angular, and irregular (Figure 3). Most of the pelves (121; 60.5%) had a curved configuration, 51 (25.5%) were angular, 19 (9.5%) were irregular, and nine (4.5%) were straight.

Of the 100 pelves examined, 51 had the same configuration of both anterior acetabular ridges and 49 had differing configurations from one side to the other.

There were no differences between genders regarding either breakdown of configurations seen (57 males and 64 females had curved configurations; p = 0.087) or side-to-side similarities or differences (29 male and 22 female pelves had the same configuration bilaterally; p = 0.161). All posterior columns were a simple hemicircle or straight classification.

Measurements

The acetabular anteversion angle was measured at the center of the acetabulum in a vertical plane to the glass table (horizontal plane) along the anterior to posterior ridges [15] (Figure 4). For acetabula that were not straight, the long axis of the anterior wall was converted to a straight line, and the midpoint then was established along that line (Figure 5).

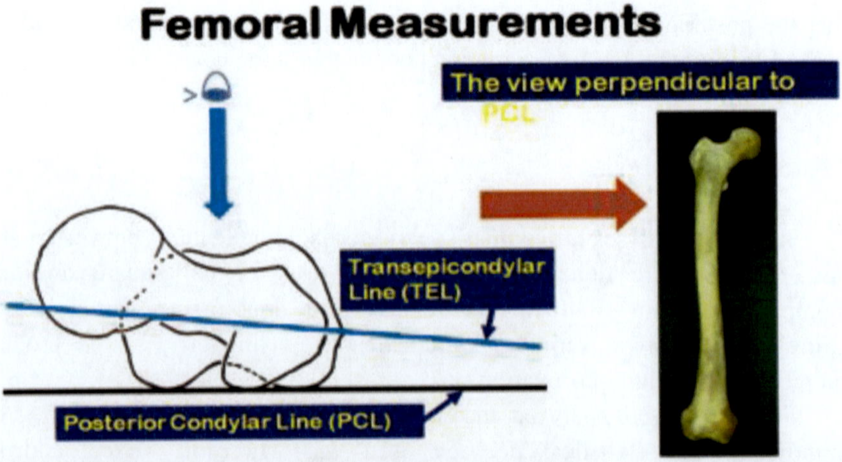

PCL: Posterior Condylar Line, TEL: Transepicondylar Line.

Figure 2. Femoral Measurements. PCL was used for definition of neutral rotation of the femur in this study. The neutral rotation may be changed to slightly internal position, however, when TEL is used for definition of the neutral rotation. Femoral measurements were performed in the view perpendicular to PCL.

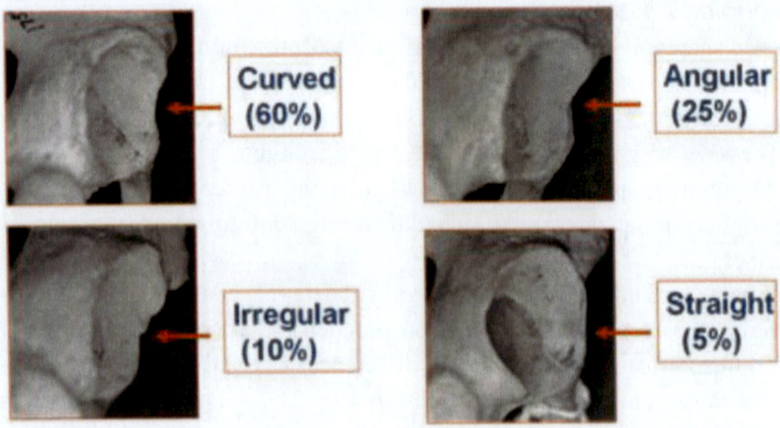

Figure 3. Anterior Acetabular Ridge Configuration was classified into four types: Curved (60%), Angular (25%), Irregular (10%), and Straight (5%). Posterior column was almost always curved or straight. There was no differences between genders as to ridge configuration. (Reproduced by permission from "Wolters Kluwer Health", Lippincott Williams & Wilkins publication: Maruyama, M., Feinberg, J. R., Capello, W. N., D'Antonio, J. A. The Frank Stinchfield Award: Morphologic features of the acetabulum and femur: anteversion angle and implant positioning. *Clin. Orthop. Relat. Res.* 393:52-65, 2001).

The notch acetabular angle was defined as an angle between a line connected the sciatic notch with the posterior acetabular ridge and a line connected both the posterior and anterior acetabular ridge. Because the line from the sciatic notch along the posterior acetabular ridge had an inclination, the notch acetabular plane is not horizontal (Figure 6).

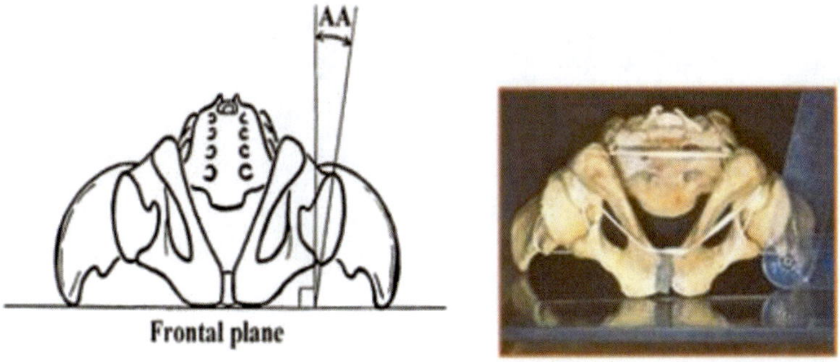

Figure 4. Acetabular Measurements: Anteversion Angle (AA). The pelvis was put on a flat glass table (craniograph) in prone position so that the anterior superior iliac spines and the pubic symphysis were contact to the glass. Thus, the glass table was regarded as the anatomic frontal plane. Acetabular anteversion angle (AA) was measured at the center of the acetabulum in a vertical plane to the glass table along the ridge.

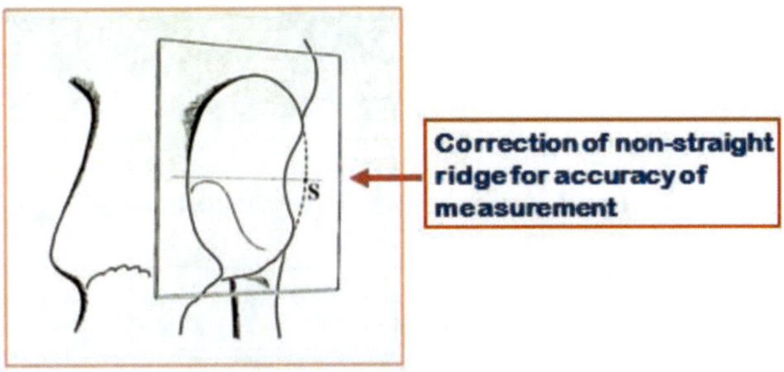

Figure 5. If correction was made in non-straight ridge for accuracy of measurement of anteversion angle, because most of the acetabulums were non-straight configuration, the anteversion angle varied (or decreased) by an average of 6.2° based on the point of measurement along the anterior ridge for the non-straight anterior ridge.

Results

The anteversion angle of the acetabulum was measured as 19.9° ± 6.6° (range, 7°-42°). The corrected acetabular angle for the acetabula that were not straight was approximately 6.2° smaller than the uncorrected acetabular angle on average (Figure 5).

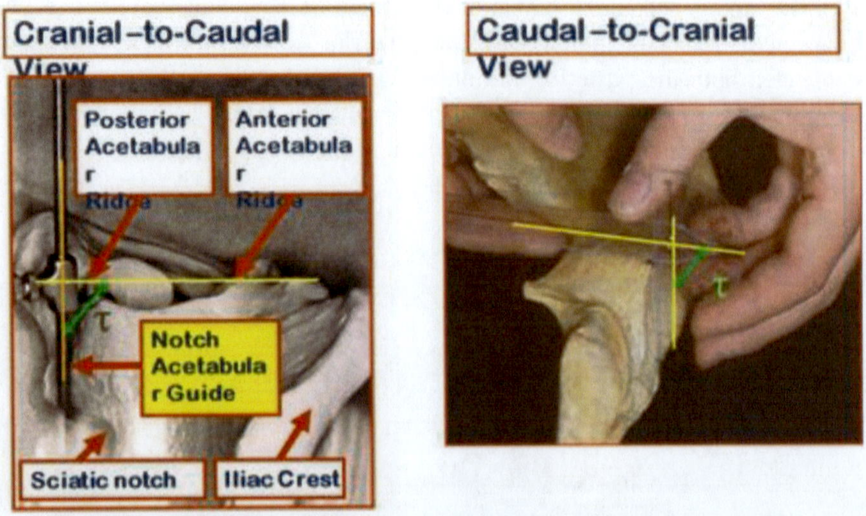

Figure 6. Notch Acetabular Angle (τ). The angle was created at the intersection of the line from the sciatic notch to the posterior acetabular ridge and the line from the posterior to anterior acetabular ridge. (Reproduced by permission from "Wolters Kluwer Health", Lippincott Williams & Wilkins publication: Maruyama, M., Feinberg, J. R., Capello, W. N., D'Antonio, J. A. The Frank Stinchfield Award: Morphologic features of the acetabulum and femur: anteversion angle and implant positioning. *Clin. Orthop. Relat. Res.* 393:52-65, 2001.)

The acetabular anteversion angle of the acetabulum was 18.5 ± 5.8° in male and 21.3 ± 7.1° in female (*p=0.002*) (Table 2).

Table 2. Acetabular Anteversion Angle

19.9° ± 6.6° (range, 7 to 42°)

- Female : 21.3° ± 7.1°
- Male : 18.5° ± 5.8°

Greater in females ($p = 0.002 < 0.05$)

The notch acetabular angle was almost perpendicular, measuring 89.0° ± 3.5° (range, 77°-100°). The angle was 88.4 ± 3.0° in male and 89.6 ± 3.8° in female (*p=0.014*). Most of the angles were ranged within 90° ± 10° (Figure 7).

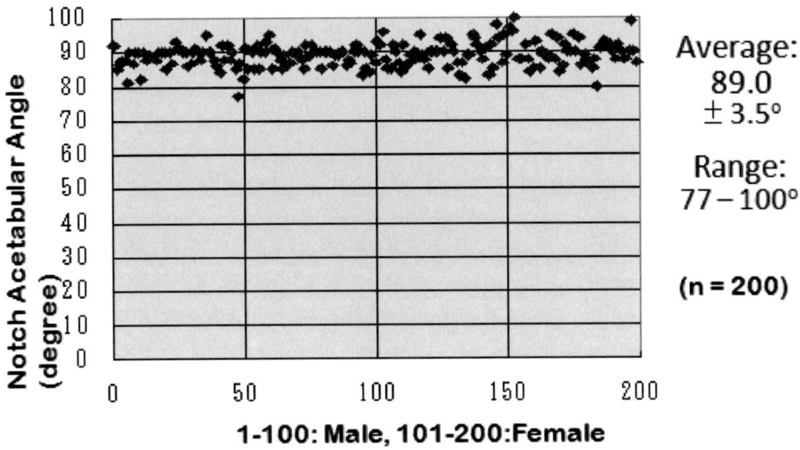

Figure 7. Notch acetabular angle was measured as an average of 89.0 ± 3.5° (range, 77-100°). The angle was 89.6 ± 3.8° in female and 88.4 ± 3.0° in male ($p = 0.014$). The angle is almost perpendicular to the face of the acetabulum and indicates anatomic anteversion.

There were no significant differences between right and left sides for males or females for any of the measured pelvic angles. There were no correlations between the notch acetabular angle and the other angles ($r^2 < 0.08$).

There were no significant differences found in the anteversion angle and the notch acetabular angle between younger (< 60 years) and older (≥ 60 years) group within each gender.

Clinical Applications

The notch acetabular angle (the angle created at the intersection of the line from the sciatic notch to the posterior acetabular ridge and the line from the posterior to anterior acetabular ridge) was measured in this study using a special tool that was placed between the sciatic notch and the posterior acetabular ridge. This notch acetabular angle has been used by two authors (JAD, WNC) to estimate the amount of acetabular anteversion necessary for accurate cup placement. With the patient in the lateral decubitus position and the surgeon on the anterior or abdominal side of the patient, the surgeon palpates using the index finger along a line from the posterior wall of the acetabulum to the greater sciatic notch.

A rod is placed parallel to the index and a mark is made on the posterosuperior aspect of the acetabulum using electrocauterization. The notch-acetabular angle, as determined in this study by direct bone measurement, had small standard deviations, such as 3.0° or 3.8°, and an extremely small range of almost 90 ± 10° in all of the cases. Although female has slightly greater angle than male, the difference was only 1.2°. For the cases with posterior osteophyte of the acetabulum, this angle was decreased but more than 80°. In addition, this angle was not correlated with the other parameters.

As a result, the notch-acetabular angle was useful to identify the anteversion of acetabulum intraoperatively. Therefore, because that line essentially is perpendicular to the plane of the face of the acetabulum, reaming for a hemispherical cup parallel to that line would result in placement of the cup in anatomic anteversion for that patient. However, if the surgeon wishes to place the acetabular component in increased anteversion, particularly when using a posterior approach, the direction of the reaming and insertion of the cup will be divergent from that line and will angle toward the top of the sciatic notch instead of the center of the iliac wing. A schematic drawing of how the desired anteversion angle is determined intraoperatively is shown in Figure 8 and 9. The fact that the notch acetabular angle was measured as nearly perpendicular in the 200 pelves in this study suggests that this angle is an accurate estimation of acetabular anteversion and that the use of this technique intraoperatively may aid the surgeon in determining more accurate placement of the acetabular component. Because an improperly angled cup is a recognized cause of dislocation, the incidence of dislocation after primary total hip arthroplasty may be reduced with use of this technique.

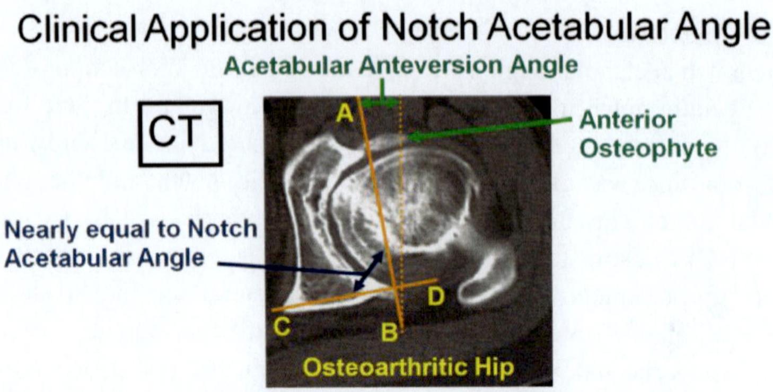

Figure 8. Notch Acetabular Angle in the Hip with osteoarthritis. Line AB indicates true acetabular anteversion. Although line AB cannot be often identified due to anterior osteophyte in case of osteoarthritic hip, line CD, which indicates nearly equal notch acetabular angle, is almost perpendicular to line AB.

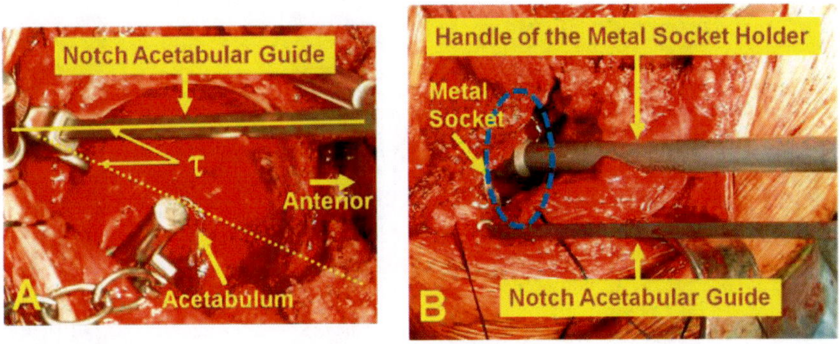

Figure 9. Surgical Application of Notch Acetabular Angle. A) Notch acetabular guide indicates the line from the sciatic notch to the posterior acetabular ridge. B) If the handle of the metal socket holder set parallel to the notch acetabular guide, the socket will be installed in the anatomically anteverted position, because the notch acetabular angle (τ) is almost perpendicular to the face of the acetabulum and indicates anatomic anteversion.

< Femur>

Morphologic Features

There were three distinct femoral shaft configurations, such as lateral, double, and medial curve, of femoral shaft in neutral rotation, in anterior-posterior view (Figure 10). Of the 200 femurs, in a neutral rotation, 162 (81 %) had lateral bowing. Thirty-two (16%) had a double or S curve, and six (3%) had medial bowing.

All double or S curves had lateral bowing proximally and medial bowing distally. With the femur in the derotated position, 124 (62%) were classified as having lateral bowing, 32 (16%) with a double curve, and 44 (22%) with medial bowing. By rotating the femur from neutral to the derotated position, the direction of bowing in 38 (19%) femurs appeared to change from lateral to medial.

In the neutral and derotated positions, femoral shaft configuration differed between males and females with the females having a greater occurrence of medial bowing (six versus 0 in neutral rotation and 29 versus 15 in derotation; $p < 0.05$). All but six of the 100 specimens had the same configuration bilaterally in the neutral position, and all but 15 had the same femoral configuration in derotation.

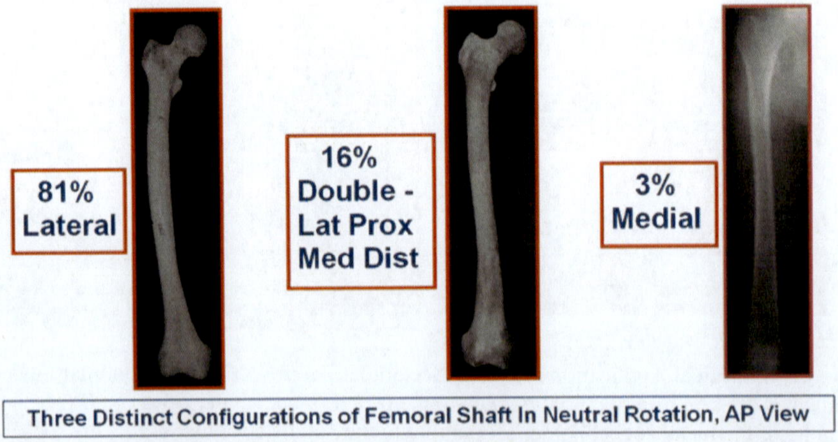

Figure 10. Femoral shaft configuration. Three distinct configurations of femoral shaft in neutral rotation, AP View, were determined. Most of the femora (81%) had the lateral bowing. The double curved femora, which were lateral proximal and medial distal, was observed in 16 %. The medial bowed femora were only 3 %. Shifting from neutral to derotation (commonly internal rotation) position, 38 (19%) femurs appeared to change from a lateral to medial bowing. (Reproduced by permission from "Wolters Kluwer Health", Lippincott Williams & Wilkins publication: Maruyama, M., Feinberg, J. R., Capello, W. N., D'Antonio, J. A. The Frank Stinchfield Award: Morphologic features of the acetabulum and femur: anteversion angle and implant positioning. *Clin. Orthop. Relat. Res.* 393:52-65, 2001).

Measurements

Femoral configuration, such as lateral bowing and anterior bowing of the femoral shaft, anteversion (or antetorsion) of the femoral neck, neck shaft angle and medial offset of the femoral head was also examined in direct measurement and in radiographs.

For consistency of measurement, the total length of the femur from the top of the greater trochanter to the base of the condyles was measured for each femur and then divided into tenths. The axis of the proximal femur (FPA) was set as a line connecting the midpoints of the mediolateral width of the femur at a distance of 2/10 and 3/10 from the top of the greater trochanter. The lateral bowing angle from the anterior-to-posterior view in neutral rotation (θ) was defined as an angle between FPA at the 2/10 and 3/10 positions and a line connecting the intersection of the neck and proximal femur axes and the center of the femoral condyles (Figure 11).

In the same manner, the lateral bowing angle from anterior to posterior in derotation was defined as the lateral bowing of the femur in the derotated position. Anterior bowing angle (δ), based on the lateral radiograph of the hip in the derotated position, was measured as the angle between the following two axes: a femoral axis at the proximal 2/10 and 3/10 lengths and a femoral axis at the 6/10 and 8/10 lengths (Figure 12).

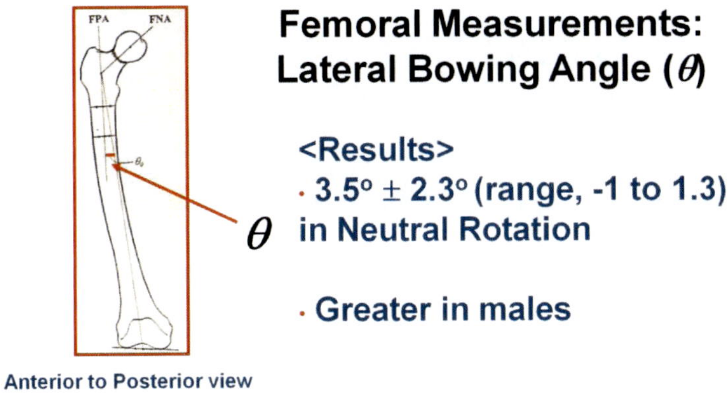

Figure 11. Lateral bowing angle (θ) was defined as an angle between proximal and long femoral shaft axis in AP view. The angle was measured as an average of $3.5^{\circ} \pm 2.3^{\circ}$ in neutral rotation and greater in males ($4.2^{\circ} \pm 2.3^{\circ}$ (range, 0 to 12) in males and $2.8^{\circ} \pm 2.1^{\circ}$ (range, -1 to 13) in females ($p = 1.5 \times 10^{-5}$). The angle was decreased in $1.5^{\circ} \pm 2.6^{\circ}$ in derotation. (Reproduced by permission from "Wolters Kluwer Health", Lippincott Williams & Wilkins publication: Maruyama, M., Feinberg, J. R., Capello, W. N., D'Antonio, J. A. The Frank Stinchfield Award: Morphologic features of the acetabulum and femur: anteversion angle and implant positioning. *Clin. Orthop. Relat. Res.* 393:52-65, 2001).

Femoral anteversion was defined based on the previous works of Billing [5] and Murphy et al. [21] The long axis of the femur (FLA) is the line defined by two points: the center of the knee (the center of the distal femoral metaphysis on a cross section through the femoral condyles), and the center of the base of the femoral neck (the center of the femoral diaphysis on a cross section through the base of the femoral neck). [14] The axis of the femoral neck (FNA) is the line defined by two points: the center of the femoral head, and the center of the base of the femoral neck. This axis bisects and passes through the midline between the anterior and posterior borders of the femoral neck. [14] In this study, neutral rotation of the femur was defined by the posterior condylar line (Figure 2). Direct bone measurement was done using a craniometer on the osteometric board for determination of the anteversion angle from the proximal (α) and distal (α_0) views (Figure 13, 14).

The reconstructed pelvis was combined with the femurs to recreate the hips so static radiographs could be obtained of all specimens. First, the right and left femurs were placed in neutral rotation and an AP radiograph was taken. Then both femurs were put in a position such that each femur was derotated (commonly internally rotated) around its long axis by the angle of anteversion from the distal view (Figure 14).

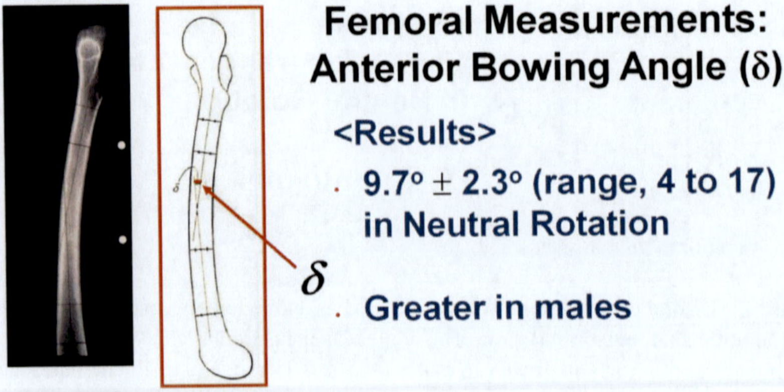

Figure 12. Femoral Measurements [Anterior Bowing Angle (δ)]: Anterior bowing angle (δ), based on the lateral radiograph of the hip in the derotated position, was measured as the angle between the following two axes: a femoral axis at the proximal 2/10 and 3/10 lengths and a femoral axis at the 6/10 and 8/10 lengths. The angle was measured as an average of $9.7^{\circ} \pm 2.3^{\circ}$ in neutral rotation and greater in males ($10.3^{\circ} \pm 2.2^{\circ}$ (range, 6 to 16) in males and $9.0^{\circ} \pm 2.4^{\circ}$ (range, 4 to 17) in females ($p = 5.7 \times 10^{-5}$). (Reproduced by permission from "Wolters Kluwer Health", Lippincott Williams & Wilkins publication: Maruyama, M., Feinberg, J. R., Capello, W. N., D'Antonio, J. A. The Frank Stinchfield Award: Morphologic features of the acetabulum and femur: anteversion angle and implant positioning. *Clin. Orthop. Relat. Res.* 393:52-65, 2001.)

Anteroposterior and lateral radiographs were taken in this derotated position. The tube was set on the pubic symphysis with a tube-to-film distance of 47.5 inches.

Radiographic magnification was corrected by comparing the actual with the radiographic diameter of the femoral head. The following dimensions were measured from the radiographs: the femoral neck shaft angle (λ) based on the AP radiograph in neutral and derotated positions (Figure 15), and the medial offset of the femoral head based on the AP radiograph in the neutral and derotated positions (Figure 16).

Femoral Measurements: Anteversion Angle in proximal view (α)

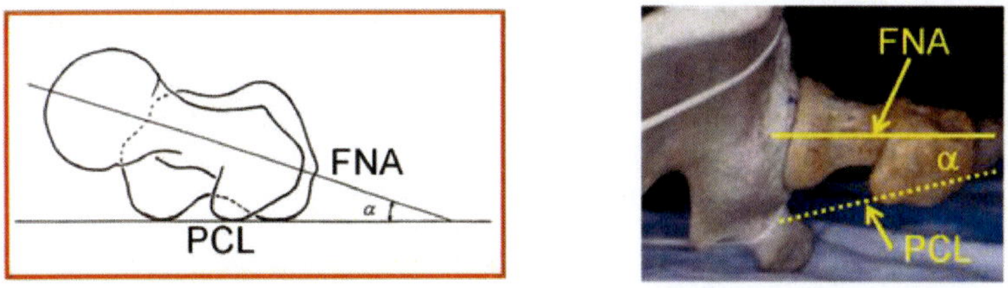

Neutral Rotation position of the femur

Internal Rotation (derotation) position of the femur and α

FNA: Femoral neck axis, PCL: Posterior condylar line.

Figure 13. Femoral Measurements. Femoral neck anteversion angle in proximal view (α) was defined as an angle between FNA and PCL. FNA was parallel to horizontal plane in derotation (commonly internal rotation (IR)) position. Shifting from neutral rotation (NR) to derotation (commonly internal rotation (IR)) position, 38 (19%) femurs appeared to change from a lateral to medial bowing.

Femoral Measurements: Anteversion Angle in distal view (α_0)

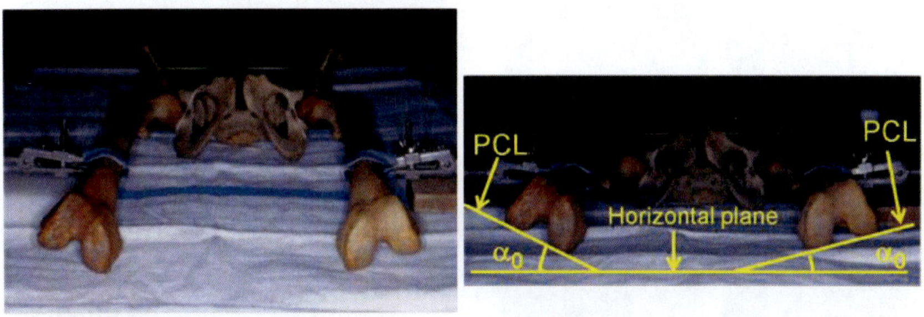

Internal Rotation (derotation) position of the femur (Distal view)

Femoral neck anteversion angle in distal view (α_0)

PCL: Posterior condylar line.

Figure 14. Femoral Measurements. Femurs were set in derotation (commonly internal rotation (IR)) position. Femoral neck anteversion angle in distal view (α_0) was defined as an angle between FNA and PCL.

Femoral Measurements:Neck Shaft Angle (λ)

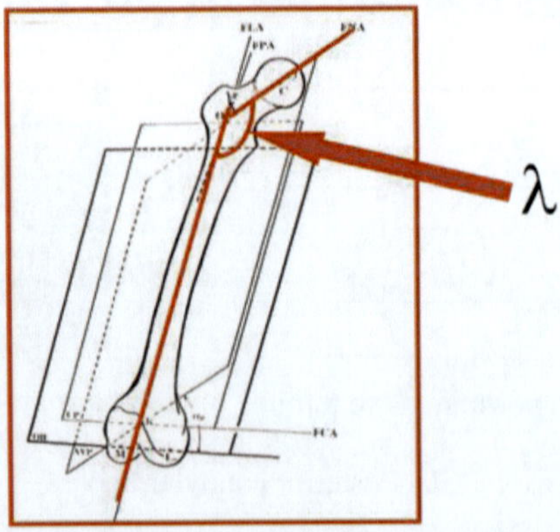

Figure 15. Femoral neck shaft angle was defined as an angle between femoral neck axis and long femoral shaft axis and measured by direct bone measurement. (Reproduced by permission from "Wolters Kluwer Health", Lippincott Williams & Wilkins publication: Maruyama, M., Feinberg, J. R., Capello, W. N., D'Antonio, J. A. The Frank Stinchfield Award: Morphologic features of the acetabulum and femur: anteversion angle and implant positioning. *Clin. Orthop. Relat. Res.* 393:52-65, 2001).

Femoral Measurements: Medial Offset

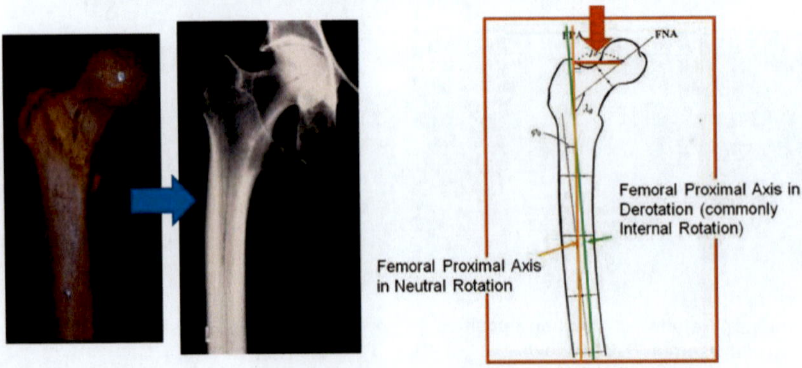

Figure 16. Medial offset of the femoral head was defined as a distance between the center of femoral head and the proximal femoral shaft axis. The offset was determined both in neutral rotated femur and in derotated femur (commonly internal rotation) on the A-P radiograph.

Results

The lateral bowing of the femoral shaft was 15.7 ± 8.7° in male and 10.1 ± 9.7° in female (*p<0.001*). If the lateral bowing angle from anterior to posterior (θ) measured ± 2° or less and the femur had minimal lateral bowing, more females than males had straight femurs (38 versus 16, respectively; *p<0.001*). The anterior bowing angle of the femur (δ) averaged 9.7° ± 2.3°, with males averaging approximately 1.3° more anterior bowing than females as measured on the lateral radiograph (*p<0.001*) (Table). The anterior bowing and lateral bowing of the femoral shaft was 10.3 ± 2.2° in male and 9.0 ± 2.4° in female (*p<0.001*). The anteversion (antetorsion) angle (α) of the femoral neck was 9.8° ± 8.5° (range, -15° to 34°) from the proximal view (9.8 ± 9.0° in male and 9.8 ± 8.0° in female (*p=0.954*)) and α_0 11.6° ± 9.1° from the distal view with no difference between males and females. There was no significant correlation between either the anterior or the lateral bowing angles and the anteversion angle (Table 3).

Table 3. Femoral Anteversion Angle

α : 9.8° ± 8.5° (range, - 15 to +34°)

- Female : 9.8° ± 8.0° (range, - 12 to +34°)
- Male : 9.8° ± 9.0° (range, - 15 to +30°)

α_0 : 11.6° ± 9.1° (range, - 30 to +34°)

- Female : 12.2° ± 7.8° (range, - 5 to +34°)
- Male : 11.1° ± 10.3° (range, - 30 to +34°)

No significant difference between males and females
No correlation with anterior or lateral bowing angles

The femoral neck shaft angle (λ) averaged 124.2° ± 6.0° in neutral rotation and 122.8° ± 5.7° in derotation (mainly internal rotation) on radiographic assessment.

The neck shaft angle of the femur in neutral rotation was 123.2 ± 6.3° in male and 125.3 ± 5.4° in female (*p=0.014*). The neck shaft angle of the femur in derotation was 121.8 ± 6.1° in male and 123.8 ± 5.2° in female (*p=0.0158*).

Females had a more valgus femoral neck-shaft angle on radiographic assessment than males in neutral and internal rotation. ($p < 0.05$) (Table 4).

The medial offset of the femoral head was significantly larger when measured in internal rotation versus neutral rotation (47.2 ± 6.1 mm versus 44.6 ± 6.7 mm, respectively; $p < 0.0001$). The medial offset of the femoral head in neutral rotation was 47.2 ± 6.4 mm in male and 41.9 ± 5.9 mm in female ($p < 0.001$). The medial offset of the femoral head in derotation was 50.1 ± 5.3 mm in male and 44.3 ± 5.4 mm in female ($p < 0.001$) (Table 5). The medial offset on the radiograph was influenced greatly by the anterolateral bowing of the proximal femoral shaft and anteversion of the femoral neck, because FPA by the bowing and FNA by the anteversion were changed in neutral and derotated position.

Table 4. Femoral Neck Shaft Angle

124.2° ± 6.0° (range, 107 to 141°) in neutral rotation

- Female : 125.3° ± 5.4°
- Male : 123.2° ± 6.3° ($p < 0.05$)

122.8° ± 5.7° (range, 106 to 140°) in derotation

- Female : 123.8° ± 5.2°
- Male : 121.8° ± 6.1° ($p < 0.05$)

Greater in females (Females had more valgus)

Discussion

Dimensional differences of the pelvis and femur between females and males were significant with a wide variation within sexes. Most of acetabula had a non-straight acetabular ridge. Configuration of the ridge cannot be detected on standard radiographs nor intraoperatively. The variations in configuration of the anterior acetabular wall and in bowing of the femur provide the information that affects the accuracy of the current methods of measurement of acetabular and femoral anteversion. The posterior acetabular ridge almost always forms a simple semicircle.

Table 5. Medical Offset of the Femur

47.2 ± 6.1 mm (range, 32 to 65 mm) in derotation
• Female : 44.3 ± 5.4 mm (range, 32 to 63 mm)
• Male : 50.1 ± 5.3 mm (range, 36 to 65 mm)
($p = 8.0 \times 10^{-13} < 0.05$)
44.6 ± 6.7 mm (range, 28.5 to 64.5 mm) in neutral rotation
• Female : 41.9 ± 5.9 mm (range, 28.5 to 61.5 mm)
• Male : 47.2 ± 6.4 mm (range, 30 to 64.5 mm)
($p = 6.2 \times 10^{-9} < 0.05$)
Greatly influenced by the anterolateral bowing of the femur as well as the anteversion of the femoral neck.

However, the anterior ridge either is curved, angular, straight, or irregularly configured (Figure 12). Because of these variations, measurement of anteversion angle is dependent upon the point of measurement along the anterior acetabular ridge. Anteversion angle could easily be over-estimated in cases of non-straight anterior acetabular ridge. In the anatomic model, the anteversion angle varied by an average of 6.2° when non-straight anterior ridges were converted to a straight configuration, therefore, the anteversion angle is dependent on the point of measurement along the ridge (Figure 3). This finding is important because the configuration of the anterior acetabular ridge is not discernible on a standard two dimensional radiograph nor is it possible to discern intraoperatively through palpation because of soft tissue, the presence of osteophyte, or both. Because greater than 95% of the 200 acetabula in this study had non-straight anterior ridges, had they undergone total hip arthroplasty, the amount of acetabular anteversion easily could have been missed to estimate in these cases depending on the point of determination along the acetabular walls intraoperatively. The presence of osteophyte, in particular those along the anterior acetabular ridge, could result in an over-estimation of the notch acetabular angle.

Medial offset measurement of the femoral head is measured based on the assumption that the long axis of the femur is straight. Because standard radiographs of the hip usually only include the proximal femur, measurements that are taken have only the proximal femur from which the long axis cannot be accurately determined. Noble et al. [23] and Preininger et al. [24] defined the medullary axis using the proximal femur as a line passing through the midpoints of the medullary canal at 20 mm proximal and distal to the canal isthmus. However, because other authors [4, 26] described changes in the medullary canal with aging and osteoporosis, the extracortical borders were used to define the femoral axes in this study. The amount of offset was influenced by rotation of the femur seen on the radiograph because of femoral anteversion with the value of the offset in derotation significantly larger than in neutral rotation. Therefore, the actual amount of offset may be underestimated on pre-op radiographic assessment. Derotation of the femur was useful in negating the effect of the femoral anteversion; however, it often is not possible to position the patient with end-stage hip arthritis who has yet to have surgery in a derotated position. Therefore, the amount of offset may be underestimated on the standard radiograph, obtained preoperatively. Medial offset also was found to be influenced by not only the anteversion of the femoral neck but also by the anterolateral bowing of the femoral shaft.

Effect of lateral bowing on the apparent long axis of femur was a change of +3.5° varus. Similarly, bowing of the femur was found to be more complex than previously thought. In addition to anterior bowing, some femurs were found to have lateral and even medial bowing. The long axis of the proximal femur was altered depending on rotation attributable to the bowing. It is known that the femur has an anterior bowing of the shaft. The current authors, however, described an average lateral bowing of approximately 13° with 81 % of femurs having lateral bowing. The effect of lateral bowing on the apparent long axis of the femur was a change toward varus of 3.5° in neutral rotation and 1.5° in the derotated position. In the current study, the lateral bowing angle was measured based on neutral rotation as determined from the posterior condylar line. However, if neutral rotation were to be determined from the transepicondylar line, then the lateral bowing angle possibly could appear decreased. Clinically, the effect of lateral bowing would mean that an implant placed in apparent neutral alignment, as determined by viewing the radiograph of the proximal femur, actually would be in slight varus regarding the true long axis of the femur in the majority of cases (Figure 9-A). It is not known whether this slight varus positioning could affect the longevity of the implant adversely although varus orientation of the stem led to increased failure with the Charnley low friction arthroplasty. [11]

Some differences were found between morphologic features of male and female pelves and femurs. The acetabular anteversion angle was greater in females, which is similar to the reported findings of others. [15, 16] No gender differences were found relative to amount of femoral anteversion, and no correlation was found between acetabular and femoral anteversion angles, which is in agreement with previous reports. [3, 28] Males had a significantly greater medial offset of the femoral head and greater anterior and lateral bowing of the femoral shaft. In neutral rotation, medial bowing of the femur was seen in six females and no males.

Proper component positioning in total hip arthroplasty relative to the amount of acetabular and femoral anteversion is important in preventing postoperative dislocation. Although this orientation is critically important, it is extremely difficult to assess intraoperatively. Even with proper acetabular component positioning, excessive anteversion or retroversion of the femoral component can lead to component dislocation. [6] Furthermore, the most common error in femoral malpositioning is excessive anteversion. [6] The most important factor in minimizing the rate of dislocation under the surgeon's control is positioning of the components in total hip arthroplasty. [18]

We previously described the study of 200 pelves and femurs with morphologic details. [14] In particular, four distinct configurations of the anterior acetabular ridge were identified and the influence of these configurations on the accuracy of anteversion angle measurement was shown. Similarly, bowing of the femur was found to be more complex than previously thought. In addition to an anterior bow, femurs were found to have a lateral bow, and in some cases and depending on rotation, this additional bow appeared to be medial.

Detailed measurements of dimensions revealed differences between males and females and between younger and older females, which may help to explain the increased incidence of dislocation after total hip arthroplasty in females. These results have implications relative to implant design in that normal anatomy has a wide variation of configurations, and therefore one implant design may not be adequate for all cases.

In particular, differences in male and female anatomy regarding anteversion angles may indicate that either different implants or different guidelines for implant positioning are needed.

Furthermore, it was shown that hip rotation influences critical measurements used in implant positioning. The use of the notch acetabular angle as an indirect estimate of acetabular anteversion may be a useful tool to decrease the incidence of dislocation after primary total hip arthroplasty.

The results of this anatomic study may explain why component positioning relative to anteversion is so difficult. On the acetabular side, the anteversion angle is influenced greatly by the point of measurement and configuration of the anterior acetabular ridge. The anteversion angle, when measured with the above considerations, differed significantly between males and females.

On the femoral side, the complexity of the bowing of the femoral shaft could influence femoral component positioning relative to anteversion and medial offset. Long femoral axis (FLA) is not differed by rotation for measurement of medial offset, it may be ideal to use this axis (FPA). The anterior and lateral bowing angles differed significantly between males and females, and these angles differed depending on the rotation of the hip.

We usually take three-dimensional CT scanning for the patients with osteoarthritis of the hip joint (Figure 17).

The next stage of this morphologic study might be evaluation of variations in configuration of normal and pathologic bone by using three-dimensional CT scanning.

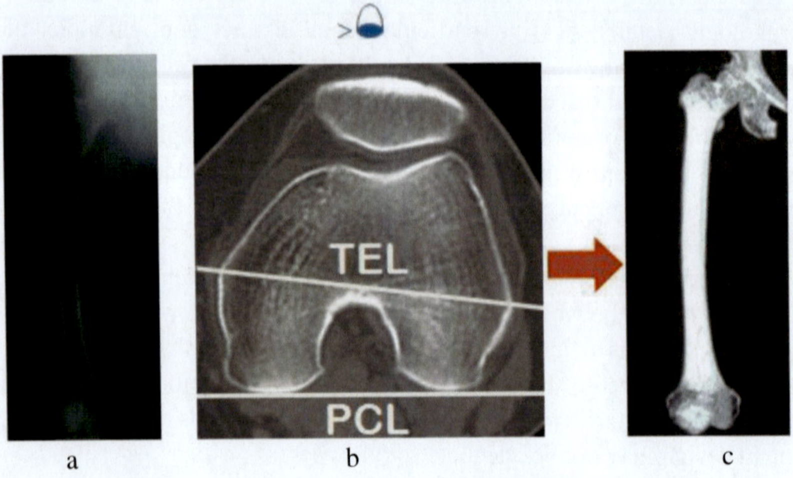

Figure 17. It is difficult to take a radiograph in neutral rotation perpendicular to PCL clinically (a). By using 3-dimensional computer tomography (3-D CT), it may be easy to evaluate femoral configuration in neutral rotation perpendicular to PCL (b and c).

Acknowledgments

The authors thank Bruce Latimer, PhD, Curator, and Lyman M. Jellema, Collections Manager, Department of Physical Anthropology, Cleveland

Museum of Natural History, for their excellent assistance and contribution for this study.

References

[1] Abel, M. F., Sutherland, O. H., Wenger, O. R., et al. Evaluation of CT scans and 3-D reformatted images for quantitative assessment of the hip. *J. Pediatr. Orthop.* 14:48-53, 1994.

[2] Ackland, M. K., Bourne, W. B., Uhthoff, H. K.: Anteversion of the acetabular cup: Measurement of angle after total hip replacement. *J. Bone Joint Surg.* 68B:409-413,1986.

[3] Anda, S., Terjesen, T., Kvistad, K. A., et al.: Acetabular angles and femoral anteversion in dysplastic hips: CT investigation. *J. Comput. Assist. Tomogr.* 15:115-120,1991.

[4] Atkinson, P. J., Woodhead, C.: The development of osteoporosis: A hypothesis based on a study of human bone structure. *Clin. Orthop.* 90: 217-228,1973.

[5] Billing, L.: Roentgen examination of the proximal femur end in children and adolescents: A standardized technique also suitable for determination of the collum-, anteversion-, and epiphyseal angles: A study of slipped epiphysis and coxa plana. *Acta Radiol.* I 10 (Suppl.): 1-80, 1954.

[6] Fackler, C. D., Poss, R.: Dislocation in total hip arthroplasty. *Clin. Orthop.* 151:169-178, 1980.

[7] Hassan, O. M., Johnston, G. H., Dust, W. N., et al.: Accuracy of intraoperative assessment of acetabular prosthesis placement. *J. Arthroplasty* 13: 80-84, 1998.

[8] Husmann, O., Rubin, P. J., Leyvaz, P. F., et al.: Three dimensional morphology of the proximal femur. *J. Arthroplasty* 12:444-450, 1997.

[9] Jaramaz, B., DiGioia, III A. M., Blackwell, M., et al.: Computer assisted measurement of cup placement in total hip replacement. *Clin. Orthop.* 354: 70-81, 1998.

[10] Khan, M. A., Brackenbury, P. H., Reynolds, I. S. R.: Dislocation following total hip arthroplasty. *J. Bone Joint Surg.* 63B:214-218, 1981.

[11] Kobayashi, S., Eftekar, N. S., Terayama, K.: Predisposing factors in fixation failure of femoral prostheses following primary Charnley low

friction arthroplasty: A 10- to 20-year followup study. *Clin. Orthop.* 306:73-83, 1994.

[12] Kristiansen, B., Jorgensen, L., Holmich, P.: Dislocation following total hip replacement. *Arch. Orthop. Trauma Surg.* 103:375-377, 1985.

[13] Lewinnek, G. E., Lewis, J. C., Taur, R., et al.: Dislocation after total hip replacement. *J. Bone Joint Surg.* 60A:217-220, 1978.

[14] Maruyama, M., Feinberg, J. R., Capello, W. N., D'Antonio, J. A. The Frank Stinchfield Award: Morphologic features of the acetabulum and femur: anteversion angle and implant positioning. *Clin. Orthop. Relat. Res.* 393:52-65, 2001.

[15] McKibbin, B.: Anatomical factors in the stability of the hip joint in the newborn. *J. Bone Joint Surg.* 52B:148-159,1970.

[16] Menke, W., Schmitz, B., Schild, H., et al.: Transversale skelettachsen der unteren extremiUit bei coxartrose. *Z Orthop. Ihre Grenzgeb.* 129: 255-259, 1991.

[17] Mensforth, R. P., Latimer, B.: Hamann-Todd collection aging studies: Osteoporosis fracture syndrome. *Am. J. Phys. Anthropo.* 180:461-479, 1989.

[18] Mohler, C. G., Collis, D. K.: *Early Complications and Their Management.* In: Callaghan, J. J., Rosenberg, A. G., Rubash, H. E. (eds). The Adult Hip. Philadelphia, Lippincott-Raven Publishers 1125-1147, 1998.

[19] Morrey, B. F.: Instability after total hip arthroplasty. *Orthop. Clin. North Am.* 23:237-248, 1992.

[20] Morrey, B. F.: Difficult complications after hip joint replacement. *Clin. Orthop.* 344: 179-187, 1997.

[21] Murphy, S. B., Simon, S. R., Kijewski, P. K., et al.: Femoral anteversion. *J. Bone Joint Surg.* 69A:169-1176, 1987.

[22] Noble, P. C., Alexander, J. W., Lindahl, L. J., et al.: The anatomic basis of femoral component design. *Clin. Orthop.* 235:148-165,1988.

[23] Noble, P. C., Box, G. G., Kamaric, E., et al.: The effect of aging on the shape of the proximal femur. *Clin. Orthop.* 316:31-44, 1995.

[24] Preininger, B., Schmorl, K., von Roth, P., Winkler, T., Matziolis, G., Perka, C., Tohtz, S. Femoral Offset (3D) in Patients without Osteoarthritis - Index Values from 200 Hip Joints. *Open Orthop. J.* 6: 578-581, 2012. doi: 10.2174/1874325001206010578. Epub. 2012 Dec. 7.

[25] Rubin, P. J., Leyvraz, P. F., Aubaniac, J. M., et al.: The morphology of the proximal femur: A three-dimensional radiographic analysis. *J. Bone Joint Surg.* 74B:28-32,1992.
[26] Ruff, C. B., Hayes, W. C.: Subperiosteal expansion and cortical remodeling of the human femur and tibia with aging. *Science* 217:945-948, 1982.
[27] Tague, R. G.: Variation in pelvic size between males and females. *Am. J. Phys. Anthropo.* 180:59-71, 1989.
[28] Visser, J. D., Jonkers, A., Hillen, B.: Hip joint measurements with computerized tomography. *J. Pediatr. Orthop.* 2:143-146,1982.
[29] Woo, R. Y. G., Morrey, B. F.: Dislocation after total hip arthroplasty. *J. Bone Joint Surg.* 64A:1295-1306, 1982.

In: Trabecular and Cortical Bone ISBN: 978-1-62808-766-6
Editor: V. Subramanium, M. Bhatnagar

Chapter VII

The Release Kinetics of Bone Morphogenetic Proteins from Demineralized Cortical Bone Grafts

William S. Pietrzak[1, 2,*] *and Saba N. Ali*[2]
[1]Biomet, Inc., Warsaw, IN, US
[2]University of Illinois at Chicago, Chicago, IL, US

Abstract

Demineralized bone matrix (DBM) grafts, created by removing the mineral phase of cortical bone by treatment with acid, are widely used throughout the skeleton for bone repair applications. Bone morphogenetic proteins (BMPs) are bioactive proteins associated with bone matrix that play an essential role in osteoinduction. Although it is believed that BMP release from DBM, in situ, is a required initial step, little is known about the release kinetics. Such knowledge, and an understanding of the parameters that affect this, can potentially lead to improved DBM graft technologies. This chapter examined the long-term release profile of BMP-7 from bovine particulate DBM (106-710 μm particle size). The

[*]Corresponding author: William S. Pietrzak, Biomet, Inc., 56 E. Bell Drive, Warsaw, IN 46581, 1-574-372-1707, 1-574-372-1683 (fax), billsp@uic.edu.

DBM (1.5 g) was exposed to 40 ml of Sorensen's buffer in closed vessels that underwent continuous mixing. Periodically, at intervals up to 84 days, the BMP-7 content of guanidine extracts of the DBM was measured with Enzyme Linked Immuno Sorbant Assay (ELISA), with ELISA used to measure the BMP-7 content of the buffer directly. The initial concentration of BMP-7 in DBM was 24.6±1.6 ng/g DBM and declined to 13.8±2.16 ng/g DBM at 84 days ($p<0.05$). However, the decline was not continuous and there was an anomalous minimum in the profile at 4-6 hours. The concentration of BMP-7 in the buffer quickly increased to 9.7±2.2 ng/g DBM at 2 hours, remained nearly constant to 24 hours, then increased in an approximately linear fashion to 146±36.7 ng/g DBM at 84 days, indicating that elution was not yet complete at this interval. Thus, there was an apparent net increase in the amount of BMP-7 contained in the system of 135.2±36.8 ng/g DBM ($p<0.001$).The results, while unusual, corroborate previously published short-term (7 day) data. A two-compartment model of the distribution of BMPs in DBM, as well as a critical examination of the guanidine-HCl extraction method, is proposed to address the unique characteristics of the elution profiles.

Introduction

Demineralized bone matrix (DBM) grafts have achieved widespread clinical application throughout the skeleton. [1] Collectively, approximately 108,000 bone graft procedures are performed per year with DBM. [1] Cleansed cadaveric cortical bone is placed into an acid medium, typically HCl solution, which solubilizes and removes the hydroxyapatite mineral phase, exposing the Type I collagen organic bone matrix. [1,2] Associated with this matrix are bone morphogenetic proteins (BMPs), especially BMPs -2, -4, and -7, which are potent inductors of endochondral bone formation. [3-9]

While it is generally thought that diffusion of BMPs from DBM is integral to the process of osteoinduction, little is known about the kinetics of BMP release. [2,10-13] Landesman and Reddi [14] suggested a soluble factor from DBM was associated with osteoinduction and showed that after an initial *in vivo* latent period, the osteoinductive potential of DBM had a half-life of 5 to 7 days. Pietrzak, et al [2], examined the change in BMP-7 concentration in bovine cortical bone during prolonged acid demineralization of up to 24 hours, finding an apparent release half-life of 26 hours in the acid medium but they did not examine BMP release from DBM under physiological conditions. The most detailed study in the literature was performed by Pietrzak, et al [15], whereby the release of BMP-7 from particulate bovine DBM into a buffer solution was measured over a 7-day period. Both the residual BMP-7 remaining in the DBM (non-eluted BMP-7) as well as its concentration in the buffer (eluted BMP-7) were measured at each time point. They found a complex elution profile that was not yet complete by 7 days, and suggested that BMPs may reside, at least conceptually, within two DBM compartments. *In vivo* assays of DBM osteoinductive potential typically place a known amount of DBM in a prepared muscle pouch of an immunocompromised mouse or rat for 28 days to allow sufficient time for new bone to form so that the response can be graded. [1,16] Whether BMP release from DBM occurs during this entire period is unknown.

Understanding the release kinetics of BMPs from DBM is important for a variety of reasons. First, many commercially-available DBM grafts consist of particulate DBM that is combined with a carrier to improve handling properties, placed in a delivery system, and packaged. [1] It is possible that during the interval between graft production and its clinical utilization a portion of the BMPs may elute into the carrier which could affect storage/stability. [13] Second, while recombinant human BMPs (rhBMP) such as rhBMP-2 and rhBMP-7 are available for clinical use, safety concerns exist regarding use of supraphysiological concentrations of these growth factors. [17] Optimal delivery has not yet been established and it is conceivable that a release profile similar to that which occurs with DBM may improve safety and efficacy of these products. Third, characterization of the BMP release profile of DBM would improve the understanding of the nature of the association of BMPs with DBM as well as that of the osteoinductive process. The purpose of this chapter was to expand previous work by investigating the long-term release kinetics of BMP-7 from DBM.

Materials and Methods

Bovine cortical long bone, which has similar composition to human bone [2], was used as a model system to avoid the stringent handling requirements of human tissue. The methods used for the preparation of DBM and quantification of BMP-7 have been previously described and are summarized below. [2,13,15,18].

Bone Processing and Demineralization

Fresh bovine cortical long bone was denuded of soft tissue, cut into cortical rings approximately 6 mm wide, and then quartered. Bone chips were created by processing the segments in water with a stainless steel industrial food disposal unit (model SS-75-27, Emerson Electric, Racine, WI). The chips were then mechanically shaken in warm water for ten minutes with a commercial mechanical paint shaker/mixer to remove any residual, adherent soft tissue. The water extract/tissue debris was discarded and the cleansed chips were lyophilized, milled to particulate form with an MF 10 Basic bone mill (IKA-Werke GMBH and Co., Staufen, Germany), and mechanically sieved through stainless steel sieves with pore sizes of 106 µm, 250 µm, 300 µm, 425 µm, 500 µm, 600 µm, and 710 µm, respectively, to determine the particle size distribution. The individual size ranges were then recombined and homogeneously mixed.

Bone particles were demineralized at room temperature in a stirred bath of 0.5N HCl for 60 minutes using a ratio of 50 ml of acid per gram of bone powder. The contents of the demineralization vessel (acid plus DBM) were then poured though a 25 µm stainless steel sieve to collect the DBM. This was followed by pouring Sorensen's buffer (SB, pH ~7.3) and then deionized water through the DBM with manual agitation for one minute, each, to remove residual acid. Excess moisture was expressed from the DBM into several layers of filter paper, followed by lyophilization. An aliquot of DBM was analyzed by digesting the organic matter and measuring the residual calcium by atomic absorption spectroscopy, using appropriate standards, to verify complete demineralization.

Elution of BMP-7 from DBM into SB

Fifty milliliter polypropylene, 30 x115 mm, conically-tipped tubes with screw caps (Falcon 352098, Bectin, Dickinson and Company, Franklin Lakes, NJ) served as the elution vessel. One-and-one-half grams of DBM and 40 ml of SB (pH~7.3, with 0.05% w/v sodium azide to prevent bacterial contamination and growth) were placed in the vessel, which was then capped (see Figure 1). The filled vessel was placed horizontally on an automated rocker plate that rotated in the horizontal plane to ensure that the vessel contents remained well-mixed. Elution was performed for intervals of 0, 2, 4, 6, 8, and 24 hours and also 3, 7, 14, 21, 28, 42, 56, and 84 days at room temperature (~25°C). Two to four independent SB extractions were performed for each elution time interval. After the prescribed time interval was reached, the vessel was held stationary in a vertical position for 5 minutes to allow the DBM to settle. The SB was then decanted into a separate polypropylene vessel and frozen at <-20°C. The DBM pellet was retrieved and excess moisture was expressed into several layers of filter paper, followed by lyophilization.

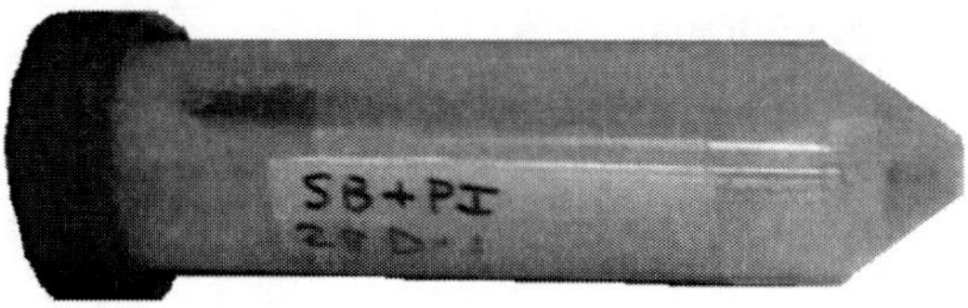

Figure 1. DBM BMP-7 extraction vessel containing 1.5 g of DBM and 40 ml of SB.

Measurement of BMP-7 Content of DBM and SB after Elution

Three-hundred milligrams of the buffer-eluted, lyophilized DBM was extracted in 5 ml of 4 M GuHCl/0.05 M Tris-HCl buffer at 4°C for 24 hours while mechanically agitated. Following centrifugation at 2,000 RPM for 5 minutes, the supernatant (SB eluent) was removed and retained. Another 5 ml of fresh guanidine solution was added to the pellet which was resuspended, followed by similar extraction, this time for 4 hours. Following centrifugation, the supernatant was removed and added to the previous aliquot, for a combined total of about 7 ml. To reconstitute the BMPs, the 7 ml of guanidine extract was placed in a dialysis tube with a molecular weight cutoff of approximately 10,000 D (part no. D9777, Sigma-Aldrich, St. Louis, MO) and dialyzed against 350 ml of 0.05M Tris-HCl buffer at 4°C for 15 hours to remove the guanidine-HCl. The Tris-HCl buffer was then replaced with fresh buffer and the dialysis was continued for another 10 hours.These dialyzed guanidine extracts were then frozen at less than -20°C.

Prior to BMP-7 assay, the guanidine extracts and SB eluents were thawed. The BMP-7 concentration of each sample was measured, in duplicate, at a wavelength of 450 nm using Enzyme Linked Immuno Sorbant Assay (ELISA) (DY-354 Duoset ELISA kit, R&D Systems, Minneapolis, MN). The concentration of BMP-7 in DBM was expressed as ng BMP-7/g DBM.The amount of BMP-7 in the 40 ml of SB was also expressed as ngBMP-7/g DBM.

Statistical Methods

The significance of differences among multiple means was determined by ANOVA and a *post hoc* Student Newman Keuls test with $p < 0.05$. Two-way comparisons were made with the unpaired Student t-test.(CD-ROM software Primer of Biostatistics, by Stanton A. Glantz, sixth edition, McGaw Hill, New York, 2005). Linear regression was performed by the method of least squares (Exel, MicroSoft Corp, Redmond, WA).

Results

Table 1 shows the size distribution of the bone particles prior to demineralization. Nearly half (46.9%) of the mass was in the range of 106 to 300 μm while the balance (53.1%) was in the range of 300 to 710 μm.Following demineralization, the residual calcium content was <0.1 wt%.

Figure 2 shows a plot of the amount of BMP-7 remaining in the DBM as well as the amount eluted into the 40 ml of SB, both expressed per g of DBM contained in the elution vessel.

Figure 3 shows a similar plot limited to the first 7 days to better illustrate the rapid initial changes that occurred. The initial (time zero) measured concentration of BMP-7 in the DBM was 24.6±1.56 ng/g. Over the first 4-6 hours, there was a significant decrease to16.4±7.04 ng/g and 17.0±6.0 ng/g, respectively ($p<0.05$). Afterward, the measured values fluctuated until 7 days and then decreased slowly until 84 days where the final value was 13.8±2.16 ng/g which was significantly less than the initial value ($p<0.05$).

Table 1. Size distribution of bone particles before demineralization

Particle size range (μm)	Weight %
106-250	35.8%
250-300	11.1%
300-425	19.9%
425-500	13.1%
500-600	10.9%
500-710	9.2%

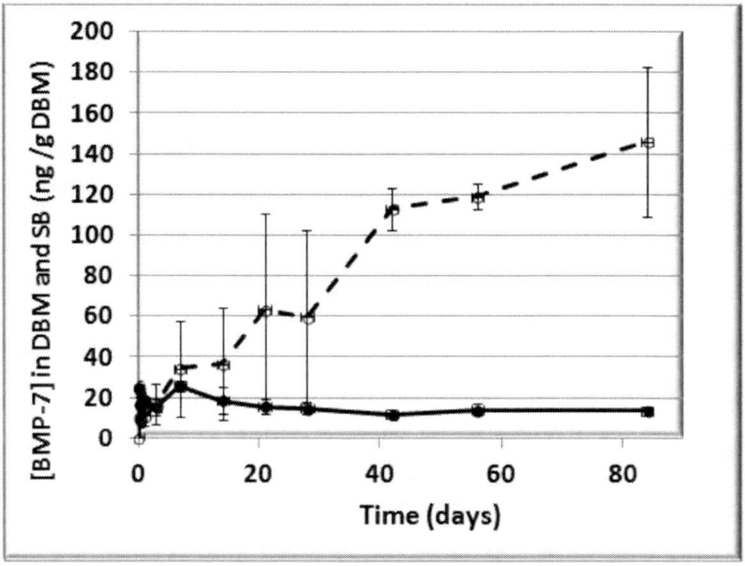

Figure 2. Amount of BMP-7 remaining in the DBM as well as in the 40 ml of SB, both expressed per g of DBM. BMP-7 in DBM: ●—●, BMP-7 in SB: ○- -○ (Ave±SD)

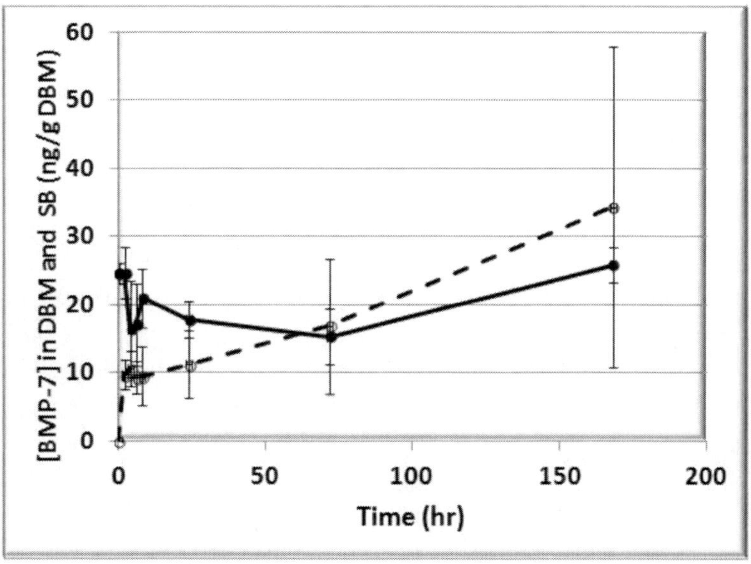

Figure 3. Amount of BMP-7 remaining in the DBM as well as in the 40 ml of SB, both expressed per g of DBM. Data shown for only first 7 days to reveal early detail. BMP-7 in DBM: ●—●, BMP-7 in SB: ○- -○. (Ave±SD).

The amount of BMP-7 in the 40 ml of SB was initially zero, then increased rapidly to 9.7±2.2 ng/g over the first 2 hours, remained essentially constant in the range of 10-11 ng/g until 24 hr, then increased in a nearly linear fashion over the balance of the 84 days. The final amount of BMP-7 contained in the 40 ml SB at 84 days was 146±36.7 ng/g. Over the course of 84 days, the measured amount of BMP-7 in the DBM declined by 10.8±2.7 ng/g while the amount in the 40 ml SB increased by 146±36.7 ng/g. Thus, there was an apparent overall increase in the amount measured BMP-7 in the closed systems of 135.2±36.8 ng/g, which was significant ($p<0.001$).

Discussion

Generally, the measurement of BMPs in DBM is performed to screen potential donors to ensure high graft osteoinductive potential [19-22] or to examine variability among commercial DBM graft preparations [23-25]. Despite the pivotal role the *in vivo* release of BMPs from DBM is believed to play in osteoinduction, these measurement techniques have not often been applied toward characterizing this release. To our knowledge, the only prior study that investigated BMP release from DBM was Pietrzak, et al [15]. They demineralized three particle size ranges of bovine cortical bone, i.e., <106 μm, 106-300 μm, and 300-710 μm, and exposed it to 40 ml SB as was done in this chapter. Initially, there was a rapid increase in the amount of BMP-7 in SB in accordance with a rapid decease in the amount remaining in the DBM, as expected. However, at 4-6 hours, the amount of BMP-7 remaining in the DBM reached an apparent minimum and then began to paradoxically increase during the balance of the 7 day study. Despite the increase in DBM BMP-7 levels after 4-6 hours, the amount of BMP-7 in the SB also continued to increase over the remainder of the study with the net effect that the total measured BMP-7 in the system (DBM and SB) was greater at 7 days than it was initially. In general, the elution profiles for the three particle size ranges were qualitatively similar.

In order to effectively compare the results of this chapter with Pietrzak, et al [15], their raw data was re-examined. As described above, they stratified the particles sizes into three ranges, i.e., < 106 μm, 106-300 μm, and 300-710 μm, and generated elution profiles for each range. This chapter utilized a single particle size range of 106-710 μm, with 46.9 wt % of 106-300 μm particles and 53.1 wt % of 300-710 μm particles. Based on these weight fractions, weighted average DBM and SB BMP-7 profiles were computed from the corresponding data of Pietrzak, et al [15], so that the comparison with this chapter would be based on the same particle size range, i.e., 106 -710 μm. These weighted average profiles are shown in Figure 4. The initial weighted DBM BMP-7 concentration was approximately 95 ng/g compared to about 25 ng/g in this chapter. It is possible that animal to animal differences between the source bone used to create the bovine DBM could account for this difference as in humans the concentration of BMP-7 can vary over a 7-fold range. [22]. Qualitatively, the weighted profiles closely resemble those for the first 7 days of this chapter, i.e., 1) The DBM BMP-7 profiles initially underwent a rapid decrease, reaching a minimum at 4-6 hours then increased by 7 days.2) The SB BMP-7 profiles initially underwent a rapid increase corresponding to the approximate time at which the DBM BMP-7 profiles reached their minimums and then continued to increase in a near-linear fashion to 7 days. So, overall, the results of this chapter corroborate those of the previous study and extend its results to 84 days.

Pietrzak, et al [15], proposed that BMPs may reside in two compartments within DBM, i.e., a "tight" and a "loose" compartment, whose names refer to the tenacity of the DBM-BMP bond. They further proposed that BMPs can only elute into the aqueous environment from the loose compartment, with the BMPs residing in the two compartments being in dynamic equilibrium with each other (Figure 5). In this way, as BMPs elute from the loose compartment to the surroundings, the loose compartment's levels are replenished from the tight compartment, which acts as a BMP reservoir. Thus, depending on the relative size of the two compartments and their associated BMP transfer rate constants, it is possible for amount of BMP in the loose compartment to reach a minimum in the elution profile. Finally, they suggested that while there is a tacit assumption that guanidine can denature and remove BMPs from DBM with 100% efficiency, it may be the case that guanidine is only effective at extracting BMPs from the loose compartment. Thus, at any point in time, guanidine would be incapable of extracting all of the BMPs associated with the DBM with those in the tight compartment being "off limits" to its effect. Over time, progressively more BMP would transfer from the tight compartment to the loose compartment and then to the aqueous surroundings so the pool of measurable BMP would increase until the tight compartment became completely depleted.

Studies that measure the amount of BMPs contained in DBM typically perform this on lyophilized DBM that has not been exposed to an aqueous medium other than that which is included in the assay, and in the absence of evidence to the contrary, this is a reasonable approach. However, before exposure to SB, we measured the concentration of BMP-7 in the bovine DBM to be 25 ng/g. After 84 days there was 14 ng/g BMP-7 remaining in the DBM and 146 ng/g in the 40 ml SB. This suggests that the DBM actually contained at least 160 ng/g of BMP-7 initially, and probably more since the SB BMP-7 profile was still increasing at 84 days.

Although there is not sufficient information to quantify the rate constants depicted in the 2-compartment model (Figure 5), one thing is apparent from Figure 2. For the interval from 21 to 84 days, there is little change in the measured amount of BMP-7 in DBM, hence, it is constant to a first approximation. During this same interval, the BMP-7 in 40 ml SB increased in a nearly linear fashion at the rate of 0.059 ng/(g·hr) (95% CI: 0.020-0.098, R^2=0.88). This would imply that the rate at which BMP-7 entered the loose compartment from the tight compartment was equal to the rate at which it left the loose compartment and entered the surrounding SB. Thus, by three weeks the early transients had disappeared and the release profile has reached what appears to be a steady state. However, it is apparent that BMP-7 release is still not complete after 84 days and, given sufficient time, the amount of BMP-7 in the SB would be expected to eventually reach a constant value while the measured value in the DBM would reach zero.

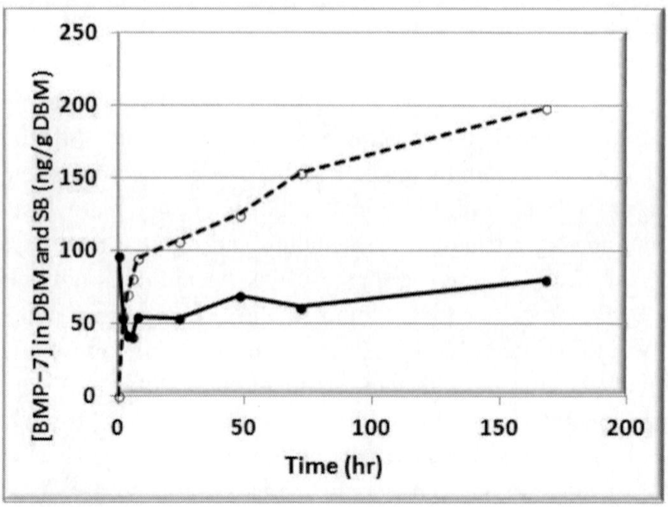

Figure 4. Weighted averages of the 106-300 μm and 300-710 μm DBM and SB BMP-7 profiles from Pietrzak, et al (15) to represent the particle size distribution of the current study. (see text for details).

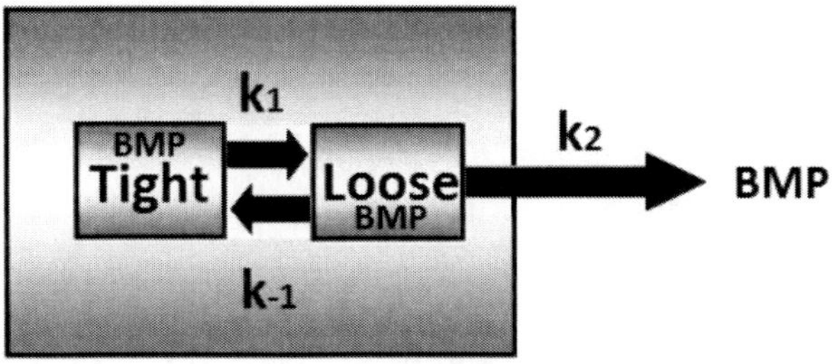

Figure 5. Schematic of the proposed 2-compartment model of the association of BMP-7 with bone matrix. k_1 and k_2 are the rate constants characteristic of the transfer of BMP-7 from the tight to the loose compartment and the elution of BMP-7 from the loose compartment to the aqueous surroundings, respectively.

As unusual as these results, and those of Pietrzak, et al [15], appear, there is some support in the literature for a two-compartment model and the influence water may have on the association of BMPs with DBM. First, although DBM has been shown by thermodynamic means to hydrate quickly in water within one minute, or so, it has been suggested that there may be longer term influences of water on the nature of the interaction of BMPs with bone matrix. [13] Second, Han, et al [26], studied the influence of elevated temperatures on the in vitro and in vivo osteoinductive activity of hydrated and anhydrous particulate DBM, finding that such activity was less stable in hydrated DBM compared to anhydrous DBM. They suggested that BMPs exist in free, loosely-bound, and tightly-bound states with DBM, further describing a reservoir concept. Third, Sampath and Reddi [27] presented evidence of a multicompartmental association of BMPs with bone, showing that some of the BMPs are even associated with the mineral phase of bone.

So, what has been learned from this chapter and why is it important? The most important thing is that the earlier work of Pietrzak, et al [15], was corroborated and extended from 7 days to 84 days, strengthening the evidence that there is a higher concentration of BMPs in DBM than the prior literature would indicate, and that DBM is capable of releasing BMPs into an aqueous environment in a timed-release fashion over a period of multiple months. Also, the in vitro duration of this release seems to be longer than would be required for the formation of new bone, much of which has been shown to occur within 28 days in animal studies.One possible rationalization for this is that as DBM is converted to new bone in vivo, the mineralization that occurs would likely impede the continued elution of BMPs. Perhaps the reservoir-like characteristic of DBM with respect to BMPs allows the maintenance of a relatively high BMP concentration gradient between the DBM and its surroundings so that despite a growing barrier to BMP release presented by the mineralization process, some release can still continue. Indeed, Han, et al [26], suggested that sustained presence of BMPs over a period of weeks may be required for the chondro/osteogenic events to occur in proper sequence. Finally, this chapter provides further support for the need to continue to develop assays so that the entire compliment of BMPs in DBM can be measured regardless of its exposure history to an aqueous environment.

The most significant limitation to this chapter was that it was performed on bovine DBM as a model system rather than human DBM. However, bovine and human cortical bone composition are very similar [2] and bovine DBM has been used previously in this type of study [15].Also, only a single type of BMP was examined, i.e., BMP-7. While it would be expected that the elution profiles of other osteoinductively-active BMPs would have similar DBM release profiles, this remains to be demonstrated. As such, future work should examine the release profile of multiple BMPs, i.e., BMP-2, BMP-4, and BMP-7, from human DBM, as well as perhaps that of other growth factors present in DBM such as TGF-β1, FGF acidic, FGF basic, IGF-I, PDGFbb, and VEGF [28,29]. The use of other non guanidine-based BMP extractants, such as 8 M urea [30], 1% sodium dodecyl sulfate [30], or collagenase [20,21] could be used to gain a better understanding of the partitioning of these growth factors among different compartments in DBM. Ultimately, this work needs to be coupled with in vivo studies to determine how these elution profiles effect the osteoinductive response in the body so that DBM grafts and possibly even rhBMP grafts can be designed with enhanced bone-forming properties.

Conclusion

The elution of BMP-7 from bovine DBM was measured over an 84 day interval. BMP-7 continued to accumulate in the buffer medium until the 84th day, showing no evidence of completion. While during this period there was an overall decline in the amount of BMP-7 remaining associated with the DBM, its profile showed an anomalous dip, or minimum, after the first 4-6 hours. Ultimately, more BMP-7 was measured in these closed systems after 84 days than was present initially. Despite the unusual nature of these results, they are supported by the literature. Future work involving a reexamination of BMP assay methods and characterization of human DBM BMP release profiles, coupled with in vivo work, will be required to more fully understand the potential implications of these results in terms of developing more effective osteoinductive bone grafts.

References

[1] Gruskin, W., Doll, B. A., Futrell, F. W., Schmitz, J. P. & Hollinger, J. O. (2012). Demineralized bone matrix in bone repair: history and use. *Adv Drug Del Rev*, *64*, 1063-1077.

[2] Pietrzak, W. S., Ali, S. N., Chitturi, D., Jacob, M. & Woodell-May, J. E. (2011). BMP depletion occurs during prolonged acid demineralization of bone: characterization and implications for graft preparation. *Cell Tissue Bank*, *12*, 81-88.

[3] Bauer, T. W. & Muschler, G. F. (2000) Bone graft materials.An overview of basic science. *Clin Orthop Rel Res*, *371*, 10-27.

[4] Reddi, A. H. (2001) Bone morphogenetic proteins: From basic science to clinical applications. *J Bone Joint Surg*, *83A*(Suppl 1), S1-S6.

[5] Sampath, T. K. & Reddi, A. H. (1984). Distribution of bone inductive proteins in mineralized and demineralized extracellular matrix. *Biochem Biophys Res Commun*, *119*, 949-954.

[6] Urist, M. R., Silverman, B. F., Buring, K., Dubuc, F. L. & Rosenberg, J. M. (1967). The bone induction principle. *Clin Orthop Rel Res*, *53*, 243-283.

[7] Urist, M. R., Iwata, H., Boyd, S. D. & Ceccotti, P. L. (1974). Observations implicating an extracellular enzyme mechanism of control of bone morphogenesis. *J Histochem Cytochem*, *22*, 88-103.

[8] Urist, M. R. & Strates, B. S. (1971). Bone morphogenetic protein. *J Dent Res*, *50*(Suppl 6), 1392-1406.

[9] Winn, S. R., Uludag, H. & Hollinger, J. O. (1999). Carrier systems for bone morphogenetic proteins. *Clin Orthop Rel Res*, *367S*, S95-S106.

[10] Urist, M. R., Granstein, R., Nogami, H., Svenson, L. & Murphy, R. (1977). Transmembrane bone morphogenesis across multiple-walled diffusion chambers: new evidence for a diffusible bone morphogenetic property. *Arch Surg*, *112*, 612-619.

[11] Urist, M. R., Dowell, T. A., Hay, P. H. & Strates, B. S. (1968) Inductive substrates for bone formation. *Clin Orthop Rel Res*, *59*, 59-96.

[12] Han, B., Yang, Z. & Nimni, M. (2008). Effect of gamma irradiation on osteoinduction associated with demineralized bone matrix. *J Orthop Res, 2008*, *26*, 75-82.

[13] Pietrzak, W. S. (2006) The hydration characteristics of demineralized and nondemineralized allograft bone: Scientific perspectives on graft function. *J Craniofac Surg*, *17*, 120-130.

[14] Landesman, R. & Reddi, A. H. (1989). In vivo analysis of the half-life of the osteoinductive potential of demineralized bone matrix using diffusion chambers. *Calcif Tissue Int*, *45*, 348-353.

[15] Pietrzak, W. S., Dow, M., Gomez, J., Soulvie, M. & Tsiagalis, G. (2012). The elution of BMP-7 from demineralized bone matrix, *Cell Tissue Bank*, *13*, 653-651.

[16] Pietrzak, W. S., Perns, S. V., Keyes, J., Woodell-May, J. & McDonald, N. M. (2005) Demineralized bone matrix graft: a scientific and clinical case study assessment. *J Foot Ankle Surg*, *44*, 345-353.

[17] Benglis, D., Wang, M. Y. & Levi, A. D. (2008). A comprehensive review of the safety profile of bone morphogenetic protein in spine surgery. *Neurosurgery*, *62*(5 suppl 2), ONS423-ONS431.

[18] Pietrzak, W. S. & Woodell-May, J. (2005). The composition of human cortical allograft bone derived from FDA/AATB-screened donors. *J Craniofac Surg*, *16*, 579-585.

[19] Traianedes, K., Russell, J. L., Edwards, J. T., Stubbs, H. A., Shanahan, I. R. & Knaack, D. (2004). Donor age and gender effects on osteoinductivity of demineralized bone matrix. *J Biomed Mater Res B Appl Biomater*, *70*, 21-29.

[20] Honsawek, S. & Dhitiseith, D. (2005). Content of bone morphogenetic protein-4 in human demineralized bone: relationship to donor age and ability to induce new bone formation. *J Med Assoc Thai*, *88*(Suppl 4), S260-265.

[21] Honsawek, S., Powers, R. M. & Wolfinbarger, L. (2005). Extractable bone morphogenetic protein and correlation with induced new bone formation in an in vivo assay in the athymic mouse model. *Cell Tissue Bank*, *6*, 13-23.

[22] Pietrzak, W. S., Woodell-May, J. & McDonald, N. (2006). Assay of bone morphogenetic protein-2, -4, and -7 in human demineralized bone matrix. *J Craniofac Surg*,*17*, 84-90.

[23] Wildemann, B., Kadow-Romacker, A., Hass, N. P. & Schmidmaier, G. (2007). Quantification of various growth factors in different

demineralized bone matrix preparations. *J Biomed Mater Res A*, *81*, 437-442.

[24] Bae, H. W., Zhao, L., Kanim, L. E., Wong, P., Delamarter, R. B. & Dawson, E. G. (2006). Intervariability and intravariability of bone morphogenetic proteins in commercially available demineralized bone matrix products. *Spine*, *31*, 1299-1306.

[25] Bae, H., Zhao, L., Zhu, D., Kanim, L. E., Wang, J. C. & Delamarter, R. B. (2010). Variability across ten production lots of a single deminmeralized bone matrix product. *J Bone Joint Surg Am*, *92*, 427-435.

[26] Han, B., Yang, Z. & Nimni, M. (2005). Effects of moisture and temperature on the osteoinductivity of demineralized bone matrix. *J Orthop Res*, *23*, 855-861.

[27] Sampath, T. K. & Reddi, A. H. (1984). Distribution of bone inductive proteins in mineralized and demineralized extracellular matrix. *Biochem Biophys Res Commun*, *119*, 949-954.

[28] Katz, J. M., Nataraj, C., Jaw, R., Deigl, E. & Bursac, P. (2009). Demineralized bone matrix as an osteoinductive biomaterial and in vitro predictors of its biological potential. *J Biomed Mater Res B Appl Biomater*, *89*, 127-134.

[29] Wildemann, B., Kadow-Romacker, A., Haas, N. P. & Schmidmaier, G. (2007). Quantification of various growth factors in different demineralized bone matrix preparation. *J Biomed Mater Res A*, *81*, 437-442.

[30] Sampath, T. K. & Reddi, A. H. (1981). Dissociative extraction and reconstitution of extracellular matrix components involved in local bone differentiation.*Proc Natl Acad Sci*, *78*, 7599-7603.

In: Trabecular and Cortical Bone
Editor: V. Subramanium, M. Bhatnagar
ISBN: 978-1-62808-766-6

Chapter VIII

Preservation of Cortical Bone in Oral and Maxillofacial Surgery Procedures

Mohammad Hosein Kalantar Motamedi
Oral and Maxillofacial Surgery, Trauma Research Center, Baqiyatallah University of Medical Sciences and Attending Surgeon, Azad University of Medical Sciences, Dental School, Tehran, Iran

Abstract

Treatment of many oral and maxillofacial lesions necessitates intraosseous access that is obtainable only at the expense of removing the maxillary or mandibular cortical, alveolar, or basal bone. Bone removal in the oral and maxillofacial region, especially in large, expansile, maxillofacial lesions, may lead to clot breakdown, facial deformity, inadequate bone formation, and delayed or insufficient healing of the defect, necessitating future reconstructive procedures with grafts [1, 3] In edentulous patients, osseous surgery may lead to loss of alveolar bone height or width, secondary surgery then being required to treat the resultant postoperative sequelae.

The composite osteoplastic flap, consisting of hard and soft tissues, is sometimes feasible with the intention of preserving the overlying bone thus, obviating the aforementioned sequelae. In this article, the applications of this flap in selected oral and maxillofacial procedures are evaluated and the advantages and clinical results are described.

Applications

In a clinical study involving 30 patients osteoplastic flaps were used to obtain intraosseous access for various indications. The patients ranged in age from 5 to 55 years (mean, 31.8 years). The technique used to obtain intraosseous access through use of the flap varied according to the site and the type of surgical procedure for that particular lesion.

Benign Expansile Intraosseous Pathosis

Cases of benign, expansile, intraosseous pathosis of the mandible, causing extensive expansion and thinning of the buccal cortex (7 cases of aneurysmal bone cysts, 3 cases of dentigerous cysts, and 1 case of central giant cell granuloma which indicated reflection and retraction of an osteoplastic flap) of the mandible were treated through use of an intraoral buccal approach.

Technique

A vertical incision is made 5 to 8 mm away from the surgical site (incision line on intact bone) and extending toward the crest (Figure 1).

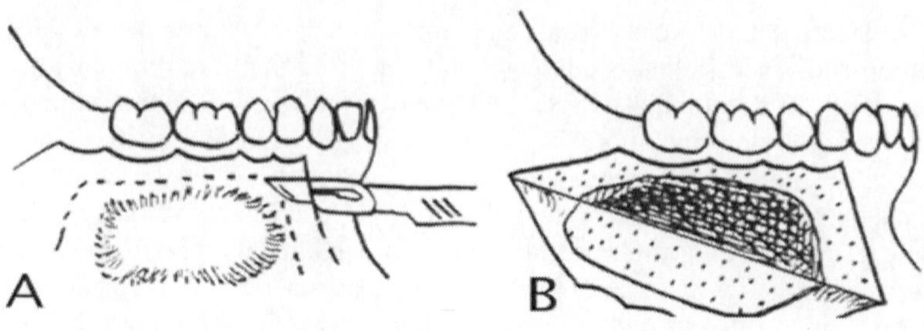

Figure 1. The osteoplastic flap technique used to access benign lesions causing cortical expansion in mandible. Margins of flap are raised (A); expanded cortex is cut anteriorly, superiorly, and posteriorly and then fractured at its base and retracted together with rest of the flap (B).

The flap is reflected 5 to 8 mm posteriorly and inferiorly until the expanded cortex is reached. The thin cortex is then cut anteriorly, superiorly, and posteriorly and reflected together with the soft tissues. In the area of the ramus, the posterior vertical incision was sometimes omitted and the bone in the posterior area was cut through a subperiosteal tunnel (depending on lesion size and on the access required). After removal of the pathologic entity, the osteoplastic flap containing the thin cortex is compressed inward. Closure of the mucosa is usually sufficient to stabilize the bone flap.

Maxillary Sinus Access

For lesions of the maxillary sinus, (4 dentigerous cysts and 1 large CGCG of the maxilla extending to the orbit in our cases) access is obtained for benign lesions by means of a vestibular incision made in the canine fossa and extending down to the periosteum several millimeters from the intended surgical site.

A supraperiosteal mucosal flap is raised toward the infraorbital nerve, and a U-shaped incision slightly larger than the size of the bony window is made in the soft tissue and periosteum over the anterior sinus wall. After elevation of the marginal periosteal soft tissues, the U-shaped bone flap is created through use of a fine bur, oscillating saw, or osteotome. The bone is cut at a bevel to prevent retraction into the antrum.

The flap is then raised and reflected pedicled on the periosteal soft tissues allowing antral access. Afterwards, the periosteal soft tissues covering the osteotomized site and the cortical bone flap are sutured. This technique has been advocated by Choung and Choung [4].

Cortical Bone Preservation in Reduction of Enlarged Maxillary Tuberosities

Preprosthetic surgical procedures are well known to practitioners; tuberoplasty is one of the most common [5, 6]. Enlarged tuberosities may be due to bony excess or a remaining third molar impaction.

In patients with bony excess, the standard procedure by which tuberoplasty is done cannot be uniformly performed for all cases. It is problematic in elderly people and those with rarefaction or osteoporosis. To prevent untoward complications or sequelae, preservation of basal bone, alveolar ridge height, maintenance of vestibular depth, prevention of deformities, irregular resorption and postoperative pain under the prosthesis is mandatory. The same may occur if bone is removed to take out an impacted third molar as it often leaves a bone defect.

Treatment of bony irregularities of the tuberosities necessitates osseous access and bone reduction obtained at the expense of removing the maxillary cortical bone.

In elderly patients, especially those with osteoporosis, removal of the thin buccal cortex of the tuberosity often leads to a "sink hole" exposing hollow rarified bone almost no marrow that may lead to irregularities under the mucosa, pain under the denture, irregular postsurgical bone resorption, alveolar deformity, and loss of denture retention. To overcome these shortcomings, we employ a composite osteoplastic flap wherein we preserve the thin overlying bone and obviate the aforementioned sequelae of tuberoplasty.

Tuberoplasty Technique

Bony Excess

Large edentulous bony maxillary tuberosities can be reduced in width with little or no reflection of the bone flap and without exposing the buccal aspect of the tuberosity. In this situation, under local anesthesia, a paracrestal palatally placed incision is made from behind the tuberosity that extends anteriorly past the undercut requiring treatment. Then, access is obtained for the removal of a wedge of crestal bone between the buccal and palatal cortices using a piezo-electric device, an electric drill and irrigation, or an osteotome and mallet. Next a triangular wedge of crestal bone is removed. Then, a buccal bone cut in the area anterior to the tuberosity is made through a subperiosteal tunnel.

Care should be taken not to encroach needlessly on the maxillary antrum. After the bone is cut vertically, the buccal cortex is in-fractured manually (Figure 2). After approximation of the wound, several sutures are placed (2-0 or 3-0 silk are sufficient to hold the flap in place). Sutures are removed 7 to 10 days postoperatively.

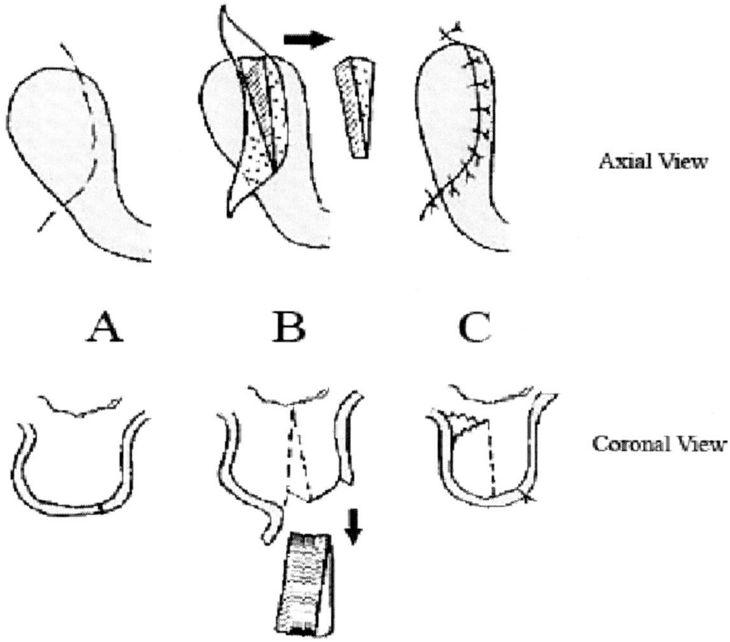

Figure 2. Technique used for reduction tuberoplasty of a large bony tuberosity. Axial view (*top*), coronal view (*bottom*).

Impacted Tooth in the Edentulous Maxillary Tuberosity

This technique may be used to access longstanding bone-impacted third molars as well as remaining roots in the edentulous tuberosity. The aforementioned technique is applied with the intent to prevent a large bony defect and loss of denture retention and alveolar crest height.

In such cases access is made via a soft tissue incision started 5 to 8 mm away from the surgical site on the alveolar crest made down to the bone, so the soft tissue margins of the flap can lie on solid bone and not on the site of the intended bone cut. The vertical incision is extended anteriorly to run along the alveolar crest beyond the impaction (Figure 3). The crestal incision should also lie palatally and not on the area of bone to be cut. A short vertical incision which is flared laterally is sometimes necessary behind the maxillary tuberosity. After the soft tissues are reflected 5 to 8 mm, the buccal cortical bone is cut vertically. The crestal bone is also cut to the necessary depth using burs, microscillating blades, or osteotomes and then split open. The bone flap is reflected outward just enough to gain access to the tooth using a periosteal elevator. The base of the flap is bent or fractured, hinging like a "trapdoor," providing access to the root or impaction.

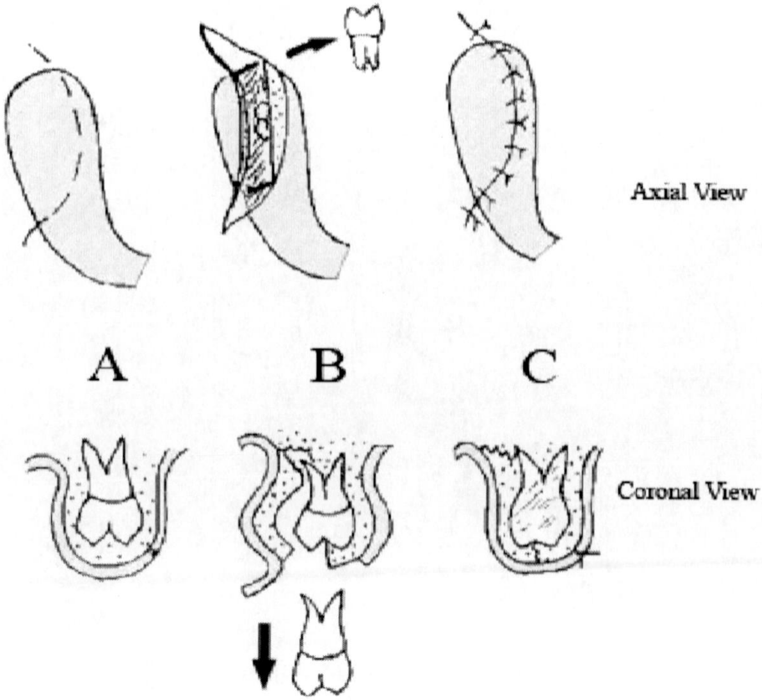

Figure 3. The osteoplastic flap procedure used for removal of impacted teeth or remaining roots in an edentulous maxillary tuberosity. Axial view (*top*), coronal view (*bottom*).

After removal of the tooth or root (with or without sectioning), the bone flap is manually adapted and pressed into place and the mucosa is sutured. Because the cortical bone overlying the tooth is not removed, the height of the alveolar bone is preserved as is the uniformity of the buccal cortices, thus, facilitating the future placement of dentures or implants. Criteria used for postoperative evaluation of the patients in this study included clinical and radiographic assessments of healing, inflammation, infection, bone formation, vestibular depth, and alveolar ridge height. Subjective evaluation of postoperative discomfort, pain, paresthesia, and so on was also done. Complete healing and bone formation were seen in each case that was treated for pathosis (the ABCs, CGCG, periapical cyst, and dentigerous cysts). Clot breakdown did not occur, and bone defects did not develop despite large cavities.

Discussion

The osteoplastic flap, first described by Wagner [7] consists of cranial bone pedicled to the temporalis muscle and aponeurosis used for craniotomy. Its application in the oral cavity was demonstrated by Ward [8] who used this flap for removal of impacted teeth with the intent to preserve the alveolar ridge. Experience with treatment of ABCs has shown the osteoplastic flap to be effective in this regard [9]. Where this lesion was large and extensive, this flap was used with an extraoral approach.

Advantages of Preserving the Buccal Cortex

Preservation of the buccal cortical plate can prevent formation of a remaining radiolucency or "surgical scar." This phenomena may occur in lesions in which the lingual cortex has resorbed and surgical removal of the buccal plate precludes postoperative bone formation in the defect, resulting in fibrosis instead of bone. In maxillary sinus surgery, preserving the wall of the sinus can be effective in preventing complications, such as oral antral fistula, facial asymmetry, and soft tissue invagination; the current opinion is that preservation of the sinus wall is preferable to its removal [10]

It should be noted that such a flap is not applicable in every case. Great care must be exercised when patients are being selected and when this flap is being attempted in proximity to the roots of the remaining dentition (so that inadvertent root exposure can be prevented). The flap is obviously contraindicated in cases involving malignant lesions or benign lesions with a high propensity to recur. There have been no recurrences in our cases during follow-up.

Advantages of Preservation of Cortical Bone in the Elderly

The osteoplastic flap described by Wilhelm Wagner (1889) [7] has had many other descriptions, such as bony mucoperiosteal flap, osteomucoperiosteal flap, periosteal bone flap, split bone technique flap, bone periosteum flap, mucoperiosteal-osteal flap, and vascularized bone flap. Its application in the oral cavity was demonstrated by Ward (1956), who used this flap for removal of impacted teeth with the intent to preserve the alveolar ridge [8].

As stated, some advocate the use of this flap in large cystic or benign expansile lesions to preserve thin expanded cortical bone, prevent clot disruption, and enhance bone formation [4,9,10].

In Tuberoplasty

Dentists and prosthodontists are commonly confronted with edentulous patients requiring full dentures that often require preprosthetic surgery, such as alveoplasty or tuberoplasty, and the edentulous patients are referred to the oral and maxillofacial surgeon. Surgery is required mainly in the area of the distobuccal prominence of the maxillary tuberosity to accommodate the buccal flange of the prosthesis.

Removal of these undercuts via flap reflection and rasping the thin cortical bone lead to a large defect with irregularities under the mucoperiosteal flap, especially in elderly osteoporotic patients. This irregular surface can be painful under a denture. Removal of the bony cortex leads to bone resorption as well. Denture retention is also compromised. As most edentulous patients are elderly, the maxilla is often rarefied. In elderly patients requiring osseous tuberoplasty, the quality of bone is usually poor, low in substance, and osteoporotic, especially in postmenopausal women. The cortical bone is also very thin. Thus, if routine tuberoplasty is performed using a bone file or bur, after only a few strokes the thin cortical bone is lost and the clinician is faced with hollow, rarefied bone with large marrow spaces, which are irregular and which break off in spicules when rasped and are impossible to smooth-out. This predisposes the patient to an uneven ridge laterally, which is painful under the denture and prone to postoperative resorption. After confronting such cases, we opted to use a bone flap to allow removal of the intercortical marrow instead of the buccal cortex, thereby decreasing the possibility of postoperative resorption and irregularities of the rarefied bone while obviating the undercuts and decreasing the bulk and width of the tuberosity. Application of this technique to large edentulous tuberosities or those encasing impactions allows for preservation of the height and width of the alveolar bone, enhances retention of dentures, and facilitates the future placement of implants.

Removal of Impacted Roots or Teeth

Use of this flap for removal of impacted roots or teeth in edentulous alveolar ridges is a simple way for the clinician to preserve the buccal bone and the existing alveolar ridge height, as well as vestibular depth. It may also be used to widen the width of the alveolar ridge to facilitate future implant placement in a thin ridge [11].

Conclusion

The osteoplastic flap is a versatile flap applicable in selected cases with benign lesions [12-16]. Use of this flap in these cases has the added advantage of maintaining the vestibular depth, because the mucogingival junction is unaltered and buccinator muscle insertions are untouched. In contrast, mucoperiosteal flaps raised for alveoplasty cause loss of vestibular depth when extended beyond the mucogingival junction.

References

[1] Peterson L Principles of management of impacted teeth In LJ Peterson, TA Indresano, RD Marciani, SM Roser (Eds.), *Principles of oral and maxillofacial surgery*, Vol. 1: JB Lippincott, Philadelphia (1992), pp. 109–122.

[2] Laskin DM. *Oral and maxillofacial surgery*, Vol. 2: Mosby, St Louis (1985), pp. 50–97.

[3] Tetsch P, Wagner W. *Operative extraction of wisdom teeth Wolfe*, London (1985), pp. 93–94.

[4] Choung PH, Choung YH. Vascularized bone flap for access to the maxillary sinus. *J Oral .Maxillofac Surg*, 55 (1997), pp. 832–835.

[5] Leonard M. The maxillary tuberosity: Indications and simple technique for reduction. *Dent Today*, 20 (2001), p. 52.

[6] Motamedi MH, Motamedi MK.Technique to Manage the Enlarged Maxillary Tuberosity in Elderly Edentulous Patients Requiring Dentures *J. Oral Maxillofac Surg*, 2011, 69:1283–1285.

[7] Buchfelder M, Ljundgren B.Wilhelm Wagner (1848-1900), 2: the osteoplastic flap. *Surg. Neurol*, 30 (1988), pp. 428–433.

[8] Ward TG/The split-bone technique for the removal of lower third molars. *Br. Dent. J.*, 47 (1956), pp. 297–301.

[9] Motamedi MH.The osteoplastic flap for access to the oral and maxillofacial region. *Oral Surg Oral Med Oral Pathol Oral Radiol Endod*, 87 (1999), p. 647.

[10] Stefansson P, Andreasson L, Jannert M.Caldwell-Luc operation: long-term results and sequelae. *Acta Otolaryngol*, 72 (1988), pp. 449–497.

[11] Kivovics P, Jáhn,M. Borbély J. Frequency and location of traumatic ulcerations following placement of complete dentures. *Int. J. Prosthodont*, 20 (2007), p. 397.

[12] Motamedi MH. Aneurysmal bone cysts of the jaws: Clinicopathological features, radiographic evaluation and treatment analysis of 17 cases. *J. Cranio-Maxillofac Surg*, 26 (1998), p. 56.

[13] Nocini P.F., Albanese M., Fior A. Implant placement in the maxillary tuberosity: The summers' technique performed with modified osteotomes. *Clin. Oral Implants Res*, 11 (3) (2000), p. 27.

[14] Motamedi MH, Navi F, Eshkevari PS, Jafari SM, Shams MG, Taheri M, Abbas FM, Motahhari P. Variable presentations of aneurysmal bone cysts of the jaws: 51 cases treated during a 30-year period. *J. Oral Maxillofac Surg.* 2008 Oct; 66(10):2098-103. doi: 10.1016/j.joms.2008.05.364. Pub Med PMID: 18848108.

[15] Motamedi MH, Stavropoulos MF. Large radiolucent lesion of the mandibular condyle. *J. Oral Maxillofac Surg*. 1997 Nov;55(11):1300-4. PubMed PMID: 9371123.

[16] Motamedi MH, Yazdi E. Aneurysmal bone cyst of the jaws: analysis of 11 cases. *J. Oral Maxillofac Surg*. 1994 May;52(5):471-5. PubMed PMID: 8169709.

Index

#

20th century, 141, 157
3D finite element model, 110
3D images, 12, 20

A

absorption spectroscopy, 194
access, xiii, 207, 208, 209, 210, 211, 215, 216
accounting, 141
acetabulum, vii, xi, xii, 164, 165, 166, 169, 170, 171, 172, 173, 174, 175, 176, 177, 178, 180, 188
acid, xii, 191, 192, 194, 203
acidic, 202
ACL reconstruction, 155
active site, 23
adaptation(s), x, 24, 29, 32, 33, 75, 115, 116, 117, 118, 120, 122, 123, 126, 128, 129, 130, 133, 134, 135, 150
adipose tissue, 17
adolescents, 40, 50, 68, 187
adults, viii, 38, 45, 57, 62, 68, 133
advancement(s), vii, 141, 156
advertisements, 140
African-American, 167
aging studies, 188
algorithm, 135
alveolar ridge, 210, 212, 213, 214, 215
alveoli, 41
amplitude, 82
anatomic site, 9, 85, 86, 87
anatomy, viii, 2, 141, 150, 151, 157, 166, 185
androgen, 26, 28
aneurysm, 151
angiogenic process, 145
anisotropy, 7, 80, 86, 106
annealing, 134
ANOVA, 196
anti-cancer, 155
antrum, 209, 210
aortic valve, 158
arthritis, 184
arthroplasty, 166, 184, 185, 188
articular cartilage, 2, 29, 31, 34, 35
articulation, 135
assessment, ix, xii, 13, 14, 23, 25, 26, 30, 65, 73, 74, 77, 79, 108, 151, 159, 165, 166, 181, 182, 184, 187, 204
asymmetry, 112
atherosclerosis, 155, 157
atherosclerotic plaque, 154
athletes, 22, 29, 121, 131
atmosphere, 41
atoms, 153
attachment, 5

automation, 151

B

b. trabecular bone (TbB), vii, 1
barium, 40, 146
barriers, 41, 150
base, 169, 176, 177, 208, 212
beams, 39
bending, 93, 97, 106, 110, 124, 153
benefits, x, 116, 121
benign, 208, 209, 213, 214, 215
bioassay, 69
biomaterials, 114, 158, 160
biomechanics, vii, 4, 18, 23, 133, 135
biopsy, 15, 148
biosphere, 41
Bisphosphonate Related Osteonecrosis of the Jaw (BRONJ), xi, 138, 152, 157
blood, 3, 9, 17, 39, 40, 56, 67, 75, 120, 158, 160
blood circulation, 40
blood flow, 56
blood supply, 9, 17, 160
blood vessels, 3, 56, 75, 158
bloodstream, 39
BMI, 151, 159
bonds, 153
bone biology, 69, 148
bone cells, 16, 17, 72
bone form, xiii, 4, 10, 11, 24, 39, 55, 56, 120, 192, 203, 204, 207, 212, 213, 214
bone growth, 3, 10, 129, 132, 134
bone marrow, vii, ix, 2, 5, 17, 22, 38, 40, 61, 69, 70
bone mass, 2, 10, 13, 14, 21, 28, 29, 59, 68, 109, 118, 131, 133
bone mineral content, 74, 99, 121
bone resorption, viii, ix, 2, 38, 55, 56, 59, 60, 62, 63, 65, 66, 67, 68, 72, 73, 210, 214
bone scan, 126
bone volume, 8, 9, 19, 20, 33, 78, 86, 93, 95, 96, 97, 99, 100, 103, 106, 108, 110, 122, 123, 124
boundary surface, 122, 123
breakdown, xiii, 169, 207, 213
breast milk, 40, 42, 54
breastfeeding, 40, 52, 54, 68
breathing, 154
bronchial asthma, 59
building blocks, 10
burn, 156

C

cadmium, 69
calcification(s), 49, 151, 158
calcium, 9, 10, 17, 18, 27, 40, 41, 52, 56, 69, 71, 72, 73, 75, 117, 142, 153, 194, 196
calcium carbonate, 153
cancer, 72, 86, 159
cartilage, 3, 4, 19, 21, 24, 25, 26, 31, 32, 35, 126
cartilaginous, 3
case study, 204
cell fate, 69
cell line, 73
ceramic, 24
challenges, 143, 144, 152
chemical, 138, 152, 157
childhood, 54
children, viii, 38, 40, 52, 53, 54, 61, 68, 136, 141, 187
China, 40
Chinese women, 71
cholelithiasis, 59
chondrocyte, 4
chromium, 10, 69
circulation, 55
classification, 169
clinical application, 20, 133, 165, 192, 203
clothing, 24
cochlear implant, 150
coefficient of variation, 96
collaboration, 167
collagen, 3, 18, 24, 31, 95, 97, 99, 108, 110, 130, 153, 192
collateral, 158

commercial, 87, 194, 198
compaction, 92
complement, 149
complexity, 88, 124, 127, 129, 152, 186
compliance, 121
complications, 188, 209, 213
composition, 32, 138, 152, 153, 156, 157, 193, 202, 204
compression, vii, 20, 34, 79, 82, 83, 84, 85, 87, 88, 89, 90, 92, 97, 100, 101, 103, 106, 108, 112, 124
compression fracture, 20, 34
computation, 89
computational modeling, vii, 123, 129
computed tomography, 12, 25, 26, 30, 86, 111, 135, 142, 150, 155, 165
computer, 65, 89, 141, 166, 186
conception, 53
condensation, 73
conference, 73, 155
configuration, xi, 164, 165, 167, 169, 170, 171, 175, 176, 182, 183, 186
confinement, 116, 120
conformity, 102
connectivity, 7, 8, 12, 79, 86, 146
conservation, 122
constituents, 5, 16, 138
consumption, 53, 68
contaminant, vii, viii, 38
contamination, viii, 38, 41, 42, 50, 67, 194
contour, 9
control group, 121, 149
convergence, 123, 126
copper, 10
correlation(s), xi, 49, 95, 97, 153, 164, 169, 173, 181, 185, 204
cortex, viii, 2, 3, 6, 9, 10, 13, 160, 208, 209, 210, 213, 214
cortical (compact) bone (CtB), vii, 1
cortical bone-resorption rate, ix, 38, 56, 57, 58, 59, 61, 62, 63, 65, 68
cortico-cancellous bone, 149
corticosteroid therapy, 59
cost, 165
covering, 209
coxa plana, 187
cracks, 79, 80, 83, 91, 93, 143
craniotomy, 213
cranium, 45
creep, 112, 113
critical period, viii, 38
critical value, 83
cross-sectional study, 27
CRP, 71
crystal structure, viii, 37
crystalline, 99
crystallinity, 99, 153, 154, 155
crystallites, 155
crystals, 18, 39, 45, 56, 68
CT, v, vii, ix, 12, 25, 77, 79, 82, 106, 107, 111, 132, 134, 141, 142, 143, 146, 147, 150, 157, 160, 161, 165, 186, 187
CT scan, 111, 142, 143, 150, 161, 165, 186, 187
cycling, 121
cyst, 212, 216
cytokines, 72

D

damping, 97
data set, 44
database, 44
deaths, 44
decay, 28, 57, 124
defects, 6, 40, 119, 152, 158, 213
deficiency, 7, 27, 28, 40, 97
deformation, 5, 40, 79, 81, 93, 97, 99, 102, 106, 108, 122, 154
degenerate, 100
degenerative joint disease, 32
dense matter, 139
dentures, 212, 214, 216
deposition, 9, 41, 70, 156
deprivation, 28
depth, 210, 211, 212, 215
detectable, 54, 56
detection, xi, 16, 49, 71, 108, 138, 139, 140, 152, 159, 165
deviation, 8

dialysis, 195
diaphyses, 5
diaphysis, 2, 3, 39, 50, 51, 177
dichotomy, 25
diet, 40, 53, 54
dietary intake, 43, 51, 52, 53, 54
diffusion, 192, 203, 204
dimorphism, 109
direct measure, xi, 82, 164, 176
direct observation, 153
discomfort, 212
discretization, 123
discrimination, 41
diseases, 21, 23, 59, 60, 62, 80, 150
dislocation, 165, 166, 174, 185
displacement, 84, 87, 88, 91, 92, 93, 94
distortions, 123
distribution, vii, xi, xiii, 1, 18, 20, 44, 75, 93, 100, 102, 106, 117, 122, 123, 124, 127, 129, 142, 155, 160, 164, 167, 192, 194, 196, 200
diversity, 142, 149
dogs, 24, 31, 120, 131, 132
donors, 86, 97, 198, 204
drawing, 174
drug treatment, 78
drugs, 19
ductility, 95, 97, 98
dysplasia, 167

E

edentulous patients, xiii, 207, 214
editors, 134
egg, 125
electromagnetic, 139
electron, 11, 15, 26, 31, 33, 35, 139, 142
electron microscopy, 11, 15, 31, 33
ELISA, xiii, 192, 195
enamel, 74
endocrinology, 17
endothelial cells, 24
energy, 12, 13, 19, 53, 81, 98, 122, 123, 139
energy density, 123
engineering, 33, 146
England, 70
environment(s), x, 41, 42, 67, 115, 116, 117, 119, 120, 122, 123, 127, 138, 139, 199, 201
environmental conditions, 117
environmental contamination, 50
environmental factors, 22
Environmental Protection Agency (EPA), 141, 157
enzymatic activity, 18
enzyme(s), 10, 203
epiphyses, 3, 5
epiphysis, 2, 26, 39, 187
equilibrium, 6, 120, 152, 199
equipment, 129
estrogen, 63, 71
ethanol, 33
ethnic background, 14
ethnicity, 14, 149
evidence, 10, 64, 132, 139, 167, 200, 201, 202, 203
evolution, 18, 21
examinations, 50, 59
excretion, 52, 68
exercise, vii, x, 26, 28, 29, 31, 35, 115, 120, 121, 122, 125, 128, 129, 132, 133, 136
exponential functions, 56, 58
exposure, ix, 14, 38, 40, 61, 72, 75, 132, 200, 202, 213
external environment, 119
extracellular matrix, 3, 99, 203, 205
extraction, xiii, 192, 195, 205, 215
extracts, xiii, 192, 195

F

facial asymmetry, 213
fat, 22, 151
FDA, 204
feces, 56
FEM, 79, 108
female rat, 27, 29
fetal development, 3, 4, 53
fetus, 40, 51, 68, 71
fiber, 153, 155, 156

fibrosis, 213
fibula, 45
finite element method, 111, 113, 114, 123
fixation, 158, 187
flexibility, vii, 1
flight, 132
fluctuations, 119
fluid, 56
fluorescence, 26, 159
fluorophores, 154
food, 40, 194
force, 5, 87, 88, 91, 92, 93, 94, 104, 121, 124, 127, 128, 129
formation, viii, 2, 3, 6, 9, 16, 17, 19, 21, 29, 54, 55, 56, 60, 65, 103, 142, 149, 201, 212, 213
formula, 19
fracture toughness, 114
fractures, viii, 2, 12, 16, 19, 20, 22, 23, 28, 29, 30, 33, 34, 78, 79, 116, 119, 141, 144, 146, 160
fragility, 19, 23, 24, 26, 80, 95, 98, 104, 109, 113, 133
fragments, 158
France, 70, 77
friction, 88, 184, 188
FTIR, 15, 153, 155, 156
FTIR spectroscopy, 155

G

gait, 128, 136
gamma rays, 139
gastrointestinal tract, 41, 52
gender differences, xi, 73, 98, 109, 163, 167, 185
gender effects, 204
genes, 25
genetic mutations, 22
genetics, 22
genotyping, 150
geometry, 72, 85, 120, 121, 122, 123, 125, 126, 133, 134
Germany, 50, 51, 194
gestation, 54
gravity, x, 115, 117, 119
growth, vii, viii, 1, 3, 4, 9, 10, 24, 28, 38, 48, 49, 50, 68, 79, 80, 84, 100, 106, 107, 110, 113, 118, 125, 129, 130, 132, 135, 155, 158, 159, 193, 194, 202, 204, 205
growth factor, 110, 155, 193, 202, 204, 205
growth hormone, 3, 24
growth spurt, viii, 38, 48
guidelines, 185

H

half-life, 41, 192, 204
hardness, 26
healing, xiii, 25, 33, 130, 131, 145, 146, 157, 160, 207, 212
health, 28, 42, 59, 62, 141, 150, 151, 160
health care, 141
health effects, 42
health status, 59
height, xiii, 49, 74, 87, 207, 210, 211, 212, 214, 215
height growth, 49
hemolytic anemia, 59
heterogeneity, 149, 153
hip arthroplasty, 165, 166, 167, 174, 183, 185, 187, 188, 189
hip joint, xi, 13, 164, 167, 186, 188
hip replacement, 166, 187, 188
histochemistry, 16
histogram, 102, 149
histology, 4, 16
history, 28, 52, 86, 133, 202, 203
homeostasis, 9, 70, 73
homogeneity, 8
hormone(s), 7, 9, 10, 11, 17, 23, 59
horses, 24, 26, 28, 29, 32, 35
human body, 40, 56, 67
human genome, 150
human subjects, 72, 155
hyaline, 4, 26
hydrostatic stress, 83
hydroxyapatite, 18, 39, 45, 56, 68, 142, 192
hyperparathyroidism, 7, 11
hypothesis, 56, 59, 187

I

iatrogenic, 16
ideal, 7, 20, 154, 186
identification, 44, 47, 134, 154
ilium, 19
image(s), 5, 6, 7, 8, 11, 13, 14, 15, 16, 17, 20, 34, 87, 138, 139, 142, 144, 150, 151, 154, 166, 187
image analysis, 8, 151
imaging modalities, 12, 15, 150
imaging systems, 15, 142
imbalances, 7, 22
immobilization, x, 28, 29, 115, 116, 117, 120, 129, 131
immune system, 72
immunocompromised, 193
impact energy, 109
impacted teeth, 212, 213, 214, 215
implant placement, 215
implants, 78, 185, 212, 214
improvements, 126
in vitro, 19, 21, 110, 159, 201, 205
in vivo, viii, 6, 14, 19, 20, 21, 23, 25, 30, 38, 44, 65, 69, 73, 134, 135, 154, 155, 157, 159, 192, 198, 201, 202, 204
incidence, 22, 72, 166, 174, 185
independence, 97, 116
indexing, 151
individuals, 6, 85, 112
induction, 203
infants, 39, 40, 75
infection, 212
inflammation, 157, 158, 212
infrared spectroscopy, 15, 153
ingestion, viii, 38, 44
inhibition, 30
initiation, ix, 10, 20, 35, 77, 78, 79, 80, 82, 83, 84, 106
injury(ies), 11, 16, 19, 22, 27, 116, 146, 151, 156, 158
inner ear, 150
insertion, 78, 174
Instron, 87
insulin, 155
integrity, xi, 137, 138, 141
intensity values, 142
interface, vii, 1, 11, 12, 20, 23
International Space Station, 120, 132
intervention, 138, 157
inventors, 142
investigative tools, xi, 138
ionizing radiation, 75, 141
ions, 18, 56
Iran, 207
irradiation, 204
irrigation, 210
ischemia, 146
isotope, 41
issues, 70, 121, 135
iteration, 123, 124

J

Japan, 163
joints, 32, 40, 120, 127

K

kinetics, vii, xiii, 69, 74, 191, 192, 193
kyphosis, 20

L

labeling, 158
lactation, viii, 38, 53, 54, 60, 68, 70, 71, 73
landscape, 152
laparotomy, 151
laws, 80, 81, 106
lead, x, xi, xiii, 7, 9, 11, 12, 20, 21, 23, 40, 54, 69, 92, 93, 115, 116, 138, 139, 146, 185, 191, 207, 210, 214
lesions, xiii, 21, 111, 207, 208, 209, 213, 214, 215
life expectancy, 21, 35
light, 4, 17, 139, 153, 154, 161
light scattering, 153
linear model, 79
liquid radioactive waste, viii, 38

liver transplantation, 158
local anesthesia, 210
localization, 90
location-dependent, viii, 2
locomotor, 31
longevity, 184
longitudinal study, 70, 73, 133
long-term retention, viii, 38
lumbar spine, 135

M

magnesium, 40
magnitude, 19, 42, 119, 121, 124, 128, 131
majority, 6, 184
mammalian skeleton, vii, 1, 72
man, 13, 56, 73
management, 27, 73, 130, 215
mandible, 148, 149, 208
manipulation, 139, 154
mapping, 110, 150, 155
marrow, 22, 75, 210, 214
mass, viii, 10, 37, 39, 68, 122, 132, 196
material porosity, 119
materials, 5, 41, 69, 70, 124, 135, 143, 203
matrix, xii, 3, 18, 24, 32, 39, 56, 83, 99, 108, 142, 144, 153, 191, 192, 201, 203, 204, 205
maxilla, 209, 214
maxillary sinus, 209, 213, 215
measurement(s), vii, viii, xii, 14, 38, 42, 43, 44, 45, 49, 52, 53, 56, 57, 58, 59, 61, 62, 65, 69, 73, 113, 117, 129, 151, 155, 164, 165, 166, 167, 170, 171, 174, 176, 177, 180, 183, 185, 186, 187, 189, 198
meat, 40
mechanical properties, ix, x, 5, 10, 12, 17, 18, 77, 78, 79, 100, 103, 106, 107, 108, 113, 115, 117, 124
mechanical testing, 91
median, 49
medical, 27, 139
medication, 19
medicine, 23, 27, 151, 159
menarche, 49, 69
menopause, ix, 2, 38, 59, 62, 63, 64, 65, 66, 67, 68, 73, 97
metabolism, 39, 42, 51, 52, 57, 59, 68, 69, 70, 71, 72, 73, 74, 154
metalloproteinase, 160
metals, 40, 70
metastatic cancer, 112
metatarsal, 35
methodology, 19, 145
mice, 155
microgravity, 120, 132
microscopy, 16, 26, 82
Microsoft, 169
microstructure(s), 23, 25, 26, 30, 34, 92, 96, 113, 131
military, 141
milligrams, 195
mineral resources, 9
mineralization, 26, 40, 104, 120, 147, 149, 150, 154, 155, 160, 201
missions, 120
mixing, xiii, 192
MMP-9, 157
model system, 193, 202
modelling, vii, 1, 3, 9, 12, 110, 112, 135
models, x, 3, 7, 12, 19, 20, 22, 79, 80, 81, 88, 96, 107, 112, 116, 122, 123, 124, 126, 127, 129, 136, 146, 149, 155
modifications, 44, 45, 117
modulus, 5, 32, 88, 89, 104, 109, 110, 111, 114
moisture, 194, 195, 205
molecular weight, 195
monohydrogen, 153
morbidity, 138
morphogenesis, 203
morphology, vii, xi, 33, 35, 95, 114, 143, 151, 161, 163, 187, 189
morphometric, 86
mortality, 72, 158
MRI, 15, 30, 142, 152
mucosa, 209, 210, 212
multiple factors, 9
muscle mass, 151
muscle stem cells, 73

muscle strength, 120
muscles, 119, 120, 127, 128, 129, 136
musculoskeletal, 116, 128, 129, 135, 142
musculoskeletal system, 116, 129

N

necrosis, 138
negative consequences, 92
negative relation, ix, 38, 61
nerve, 209
neutral, xii, 18, 125, 164, 170, 175, 176, 177, 178, 179, 180, 181, 182, 184, 185, 186
New England, 23, 26, 35
New Zealand, 33, 35
NIR, 154
Nobel Prize, 139
nodes, 19, 122, 124, 126
non-linear analysis, ix, 77, 106
normal aging, 107
normal development, 40
North America, 25, 159
nuclear weapons, 41
nutrient(s), 10, 144
nutrition, 9, 22
nutritional status, vii, 1, 9

O

objectivity, 84
optical properties, 157
optimization, 122, 123, 124, 126, 127, 135, 136, 150
optimization method, 122, 123, 127
oral cavity, 213, 214
orbit, 209
orchidectomy, 29
organ(s), 9, 17, 21, 78, 80, 96, 97, 107
organic matter, 194
ossification, 3, 4, 24, 154, 155, 156, 160
osteoarthritis, 6, 12, 21, 22, 23, 25, 26, 29, 30, 31, 32, 35, 40, 166, 167, 174, 186
osteocyte, 16
osteogenesis, 145, 157
osteogenesis imperfecta, 148, 167
osteomalacia, 148
osteoporosis, viii, 2, 6, 7, 10, 14, 21, 23, 24, 25, 27, 71, 72, 73, 78, 80, 97, 112, 133, 136, 148, 159, 184, 187, 209, 210
osteotomy, 158
outpatient, 59
overlap, 147

P

pain, 27, 210, 212
palpation, 183
parallel, 8, 17, 18, 173, 174, 175, 179
patella, 135
pathogenesis, 26, 34, 35, 72
pathology, viii, 2, 15, 62, 151, 159
pathophysiology, 27, 34
pelvis, vii, xi, xii, 1, 135, 163, 164, 165, 166, 167, 168, 171, 178, 182
peptide, 157
perforation, 8
perfusion, 146, 147
perinatal, 73
periosteum, 3, 4, 61, 69, 209, 214
permission, 11, 16, 170, 172, 176, 177, 178, 180
permit, 143
petroleum, 134
pH, 18, 194
pharmaceutical, 119, 120, 121
phosphate(s), 18, 56, 153
phosphorus, 10, 117
physical activity, 22
physical properties, 153
physicians, 151, 156
physio-anatomy, viii, 2
physiology, xi, 131, 133, 137, 157
physiopathology, 14
pigs, 40, 120
placenta, 41
plants, 41
plasticity, 39
playing, 12

plutonium, 40, 41
pneumonectomy, 130
Poisson ratio, 89, 104
polymethylmethacrylate, 31
polypropylene, 194
population, 14, 21, 23, 25, 41, 42, 61, 66, 70, 71, 72, 73, 151, 167
porosity, 7, 27, 109, 117, 120
positive correlation, 82
posterior acetabular ridge, xi, 164, 171, 172, 173, 175, 183
pregnancy, viii, 38, 40, 51, 53, 54, 60, 68, 70, 71, 73
preparation, 193, 203, 205
preservation, vii, 158, 209, 213, 214
prevention, 121, 152, 210
principles, vii, 2, 17
probability, 20
probe, 159
progenitor cells, 75
propagation, ix, 77, 78, 79, 80, 82, 83, 84, 85, 92, 93, 104
prostate cancer, 28
prostheses, 165, 187
prosthesis, 187, 210, 214
protection, 18, 71, 138, 155
proteins, vii, xiii, 191, 192, 203, 205
proteolytic enzyme, 55
puberty, 49, 50, 70
public health, 140
publishing, 27

Q

quantification, 16, 30, 193

R

race, 167
racehorses, 30
radiation, ix, 12, 14, 15, 30, 37, 38, 40, 41, 42, 61, 69, 72, 75, 139, 141, 148, 149, 150, 153, 155
radiation damage, 155
radioactive contamination, 41
radioactive isotopes, 40, 41, 67
radioactive tracer, 55, 65
radioactive waste, viii, 38
radiodensities, xi, 137
radiography, 30, 142, 154
radiotherapy, 149, 157
radium, 40
radius, 14, 23, 27, 30, 66, 84, 111, 132
radon, 142
Raman spectra, 153
Raman spectroscopy, xi, 138, 152, 153, 155, 156, 157
reality, 138
recommendations, 116
reconstruction, 42, 70, 90, 142, 167
recycling, 56, 57
redistribution, 83
regeneration, 69
registry, 44
regression, 196
relative size, 199
relevance, 74, 79, 146, 147
remodelling, vii, 1, 2, 3, 4, 7, 9, 12, 17, 19, 21, 26, 27, 30, 131, 134, 158
repair, xiii, 4, 9, 16, 18, 19, 20, 21, 25, 33, 73, 117, 119, 141, 144, 151, 152, 191, 203
reprocessing, 41
reproduction, 126
reproductive age, viii, 38, 62, 65
requirements, 3, 51, 68, 122, 125, 143, 193
researchers, 142, 156
resistance, 5, 18, 20, 81, 104, 113, 121, 127
resolution, 6, 7, 12, 14, 20, 30, 31, 34, 79, 86, 88, 111, 112, 142, 143, 144, 146, 148, 150, 159, 160
response, 5, 6, 11, 16, 18, 24, 29, 31, 35, 61, 118, 119, 122, 124, 126, 132, 133, 134, 193, 202
responsiveness, 138
restoration, 54
restrictions, 122, 154
rheumatoid arthritis, 59
rickets, 40

rings, 194
risk(s), x, 7, 13, 14, 15, 32, 79, 115, 116, 118, 119, 120, 147, 151, 159, 166
risk assessment, 7, 79
risk factors, 166
rodents, 25, 26, 155
rods, 5, 7, 8
room temperature, 87, 194, 195
root(s), 211, 212, 213, 215
rotations, 81
routines, 125
rubber, 5, 168
rules, 75
Russia, viii, 38, 41, 50, 68

S

safety, 193, 204
sarcopenia, 151
scanning electron microscopy, 16, 17, 26, 33
scapula, 39
scattering, 142
science, 17, 203
sclerosis, 6, 11, 12, 22, 26
seafood, 40
sediments, 42
semicircle, 183
sensitivity, 85, 89, 104, 108, 154
serum, 71
settlements, 42, 44, 49, 58, 75
sex, 7, 9, 10, 11, 14, 22, 23, 71, 75
sex hormones, 10, 11, 22
sex steroid, 71
shape, x, xi, 5, 7, 91, 115, 117, 118, 119, 121, 122, 123, 124, 126, 127, 129, 135, 164, 165, 169, 188
shear, 93, 101
sheep, 24, 120
shores, 42
showing, 4, 50, 96, 201, 202
Siberia, 40
side effects, 119
signals, 131
signs, 21, 108, 151
silicon, 69
silk, 210
simulation(s), vii, x, 78, 79, 80, 84, 88, 89, 104, 105, 106, 107, 108, 113, 123, 125, 126, 129, 130, 161
skeleton, vii, viii, x, xiii, 1, 2, 10, 18, 21, 23, 24, 37, 39, 40, 43, 45, 49, 50, 52, 54, 55, 68, 71, 72, 73, 137, 152, 191, 192
skin, 155
smoking, 22
smoothing, 123
soccer, 121
sodium, 194, 202
software, 169, 196
solution, 56, 87, 93, 122, 123, 124, 192, 193, 195
South Africa, 31
Soyuz, 120
spastic, 136
species, 18, 21, 121
specific surface, 39
spectroscopy, 153, 155
spine, 14, 109, 126, 133, 204
spine fractures, 109
spongy bone, vii, 1, 111
stability, 118, 158, 188, 193
standard deviation, xii, 14, 51, 64, 165, 174
standard error, 57
state(s), viii, 2, 74, 83, 101, 120, 122, 123, 126, 129, 157, 200, 201
statistics, 86
steel, 131, 132, 194
sternum, 39, 44, 45
storage, viii, 38, 45, 119, 193
stress fracture, 118, 131
stress-strain curves, 98
stretching, 123, 153
strontium, viii, 38, 40, 41, 42, 45, 52, 55, 56, 57, 67, 69, 70, 71, 72, 73, 74
structural changes, 143
structural characteristics, 14
structural formation, 39
structure, viii, 2, 3, 4, 7, 8, 10, 12, 14, 17, 19, 20, 25, 32, 39, 81, 86, 100, 107, 109,

117, 118, 119, 126, 130, 131, 138, 143, 151, 153, 165, 187
style, 66
subchondral bone, vii, 1, 2, 4, 6, 24, 25, 26, 28, 29, 31, 32, 34, 35
subdomains, 19
substitution, 99
substrates, 154, 203
sulfate, 146, 202
superimposition, 142
surface area, vii, 2, 9, 23, 67, 86
surgical removal, 213
symptoms, 40
syndrome, 188
synthesis, 130
torsion, 124
toxic effect, 40
toxicology, 67
trace elements, 68
training, 11, 26, 29, 31, 32, 35, 121, 132
translation, 152
transmission electron microscopy, 16
transport, 144
trauma, xi, 20, 138, 141, 151, 167
treatment, xii, 21, 33, 72, 78, 80, 138, 150, 151, 191, 210, 213, 216
trochanter, 5, 85, 86, 176
tuberculosis, 140
turnover, 21, 34, 50, 52, 54, 67, 68, 71, 73, 149, 154

T

Tajikistan, 40
target, 9, 17, 117, 119, 121, 129
techniques, x, 16, 82, 116, 122, 123, 127, 142, 153, 155, 159, 198
technology(ies), xi, 16, 138, 139, 141, 142, 143, 152, 165
teeth, 40, 43, 74, 215
temperature, 87, 195, 205
tendon, 5, 29, 127
tension, 79, 82, 83, 84, 85, 88, 89, 100, 101, 107, 158, 159
testing, 41, 87
tetracycline antibiotics, 159
tetracyclines, 154
texture, 32
TGF, 202
therapeutic agents, 12
therapy, 28, 59, 147
thinning, viii, 2, 9, 21, 97, 148, 149, 208
third molar(s), 209, 210, 211, 216
thorax, 135
three-dimensional model, 123
tibia, 6, 7, 14, 30, 39, 65, 66, 114, 133, 160, 189
time periods, 44, 50, 54, 63
tissue engineering, 145, 152
tooth, 212

U

UK, 27, 43
Ukraine, 41
ulna, 25
ultimate bone strength, vii, 1
ultrastructure, 143, 156, 160
unification, 6
uniform, 96, 124
unions, 146
United Nations, 75
United States (USA), 27, 167
universality, 129
uranium, 40
urea, 202
urine, 56, 120
UV, 139

V

vacuum, 139
valgus, 166, 182
validation, 79, 85, 107, 110, 155
variables, 7, 8, 15, 79, 169
variations, 14, 106, 109, 112, 122, 126, 127, 165, 182, 183, 186
varus, xii, 165, 184
vascular system, 160

vasculature, 146
vegetables, 40
velocity, 87
vertebrae, vii, 1, 2, 19, 20, 39, 78, 82
vertebrates, 21
vessels, xiii, 144, 192
viscosity, 80, 106
visualization, 143, 157

W

walking, 121, 136, 168
water, 41, 42, 50, 67, 68, 194, 201
weakness, 20, 136
wealth, 149, 150, 152
weapons, 41
weight ratio, 39
welfare, 21
wild type, 155
withdrawal, 32
worldwide, 41

Y

yield, 19, 24, 79, 82, 83, 84, 87, 88, 91, 95, 100, 103, 108, 113, 124, 129, 151
young women, 28

Z

zinc, 10